Reconfiguring Truth

Reconfiguring Truth

Postmodernism, Science Studies, and the Search for a New Model of Knowledge

STEVEN C. WARD

ROWMAN & LITTLEFIELD PUBLISHERS, INC.
Lanham • Boulder • New York • London

ROWMAN & LITTLEFIELD PUBLISHERS, INC.

Published in the United States of America
by Rowman & Littlefield Publishers, Inc.
4720 Boston Way, Lanham, Maryland 20706

3 Henrietta Street
London WC2E 8LU, England

British Cataloging in Publication Information Available

Library of Congress Cataloging-in-Publication Data

Ward, Steven C.
Reconfiguring truth : postmodernism, science studies, and the search for a new model of knowledge / by Steven C. Ward.
p. cm.
Includes bibliographical references and index.
ISBN 0–8476–8259–5 (cloth : alk. paper). — ISBN 0–8476–8260–9 (paper : alk. paper)
1. Knowledge, Theory of. 2. Truth 3. Relativity. 4. Postmodernism. 5. Social realism. 6. Social sciences—Philosophy. 7. Humanities—Philosophy. 8. Science—Philosophy. I. Title.
BD175 W37 IN PROCESS 1996
149—dc20 96–17631
CIP

ISBN 0–8476–8259–5 (cloth : alk. paper)
ISBN 0–8476–8260–9 (paper : alk. paper)

Printed in the United States of America

™ The paper used in this publication meets the minimum requirements of American National Standard for Information Sciences—Permanence of Paper for Printed Library Materials, ANSI Z39.48–1984.

To my mother and the memory of my father

Contents

Preface

It appears that the hostility and dislike between the humanities and sciences described by C. P. Snow in the late 1950s has now developed into a full-fledged war over the status of knowledge. As is the case with most wars, it is often difficult to differentiate the aggressor from the victim. On the one hand, scientific realists claim that science is under siege by an unholy alliance of feminists, multiculturalists, literary and cultural theorists, and sociologists of science. They contend that rationality and science are threatened by this motley group of postmodernists who want relativism to replace realism and politics to supplant reason. On the other hand, postmodernists claim that they are only illustrating the human, all too human, characteristic of all knowledge. It is scientists who are guilty of denying the political and cultural situatedness of their own knowledge form.

This book is an investigation of some of the recent attempts to reconfigure truth in the humanities and social sciences and how these attempts are connected with the current knowledge wars in academia. It is also an exploration of the implications of actor-network theory found in social studies of science for the debate between scientific realists and postmodern textualists. In this work I examine four contemporary approaches to knowledge: scientific realism, postmodern textualism, social realism, and actor-network theory. The first three chapters survey the modern way of knowing and various postmodern attempts to deconstruct the foundational and representational claims of modernists. Chapters 4 and 5 explore the doctrine of social realism and some of its limitations and paradoxes. The remaining chapters examine actor-network theory and its relation to the knowledge wars and the debate between modernists and postmodernists. While this book makes no claim to solve the modern/postmodern impasse, I hope that it will at least begin the process of reframing some of the questions and issues.

Acknowledgments

Acknowledgments often have a long list of institutional supports usually accompanied by cryptic grant numbers or gracious recognitions of leisurely leaves of absences spent in exotic research locations. I wish I could acknowledge such contributions, for they would have made the task much easier. However, as is often the case with theoretical works of this type, I generally only have my colleagues and friends to acknowledge. First, I must thank the following people who read and made suggestions on an earlier version of the manuscript: Stephan Fuchs, Diane Henningfeld, Steve Sweet, and Marcy May. Their comments and suggestions have greatly enhanced the book. I am also extremely grateful to Marilyn Kuruz and Deirdre Souchere for their editorial and production assistance. Also, thanks to my students at both Adrian College and Western Connecticut State University who have always forced me to clarify my ideas. Finally, thanks to Karen Roth for Michigan and other noble compromises over the last ten years.

Parts of some chapters have been taken from two earlier publications, "In the Shadow of the Deconstructed Metanarratives: Baudrillard, Latour and the End of Realist Epistemology" (*History of the Human Sciences* 7 (4): 73-94) and "The Revenge of the Humanities: Reality, Rhetoric, and the Politics of Postmodernism" (*Sociological Perspectives* 38 (2): 109-128). I am grateful for permission to use parts of these articles here.

Introduction

Postmodernism and its conceptual corollaries--deconstruction, intertextuality, discourse, rhetoric, and hyperreality-- have become the intellectual and cultural buzz words of the 1980s and 1990s. Such popularization and proliferation, however, have not resulted in an improved understanding of what postmodernism is about. For instance, in the last fifteen years the term has been used to describe television commercials, urban skyscrapers, feminist theory, poststructuralism, and the gay and lesbian rights movement. Once we clear away the variety of uses, however, it appears that at the center of the debates surrounding postmodernism, at least many of those occurring in the humanities and the social sciences, lie important and long-standing questions concerning the possibility of obtaining objective, context-independent truth.[1] All varieties of postmodernism are firmly antiepistemological. Postmodernists distrust all modern attempts to ground representation and knowledge in some philosophical or theoretical project, such as the Kantian subject, the perfection of language, or the empirical methods of science (see Murphy 1990).

Within the last ten years the antiepistemological stance of postmodernism has created a sea of controversy, debate, and confrontation. Critics see it as giving rise to a contemporary crisis of representation—of announcing the end of all fixed meaning and of collapsing the epistemological distinctions between belief and knowledge, science and literature, authenticity and simulacrum, and, ultimately, truth and falsehood (see Habermas 1981, 1987; Gross and Levitt 1994; O'Neill 1995). For these critics, postmodernism's leveling of all epistemic distinctions prohibits knowledge practitioners from providing any accurate representations of the real and, as a consequence, any type of definitive truth claim.

Politically it also inhibits theorists from offering any guide for the emancipation of oppressed groups. As such, postmodernism's antiepistemological stance leads contemporary thought into a nihilistic, antifoundational hermeneutics where perspectivism runs rampant, interpretation piles on top of interpretation, "anything goes," and progressive political action is impossible. On the other hand, proponents of postmodernism claim that it marks an end to the hegemonic, Eurocentric, androcentric, and intellectually confining philosophic system of modernity (see Ward 1996, 13). It therefore marks an important and, for some, an inevitable turning point in the development of postpositivist and multicultural approaches to knowledge and truth (see Hutcheon 1988; Turner 1994).

As the term "postmodern" has come into widespread use and the conflicts over its central propositions have escalated, theorists in some circles have begun to view the issues surrounding postmodernism and its corollaries as becoming increasingly tiring and practically self-defeating. Some have become weary of "reading" and "deconstructing" their way to obliteration or of searching tirelessly for the infinitely reducible voice of the "marginalized other." Some, particularly in history and the social sciences, have begun to long for the good old days of "normal" science, when they could treat things as "normal" research objects with "normal" research methods without running the risk of immediate deconstruction. This "turning back" also seems problematic, however. The questions and dilemmas opened up by postmodern theory seem impossible to ignore or set aside. A nonproblematic return to an intellectual "business as usual," where representations and objects naturally correspond, where professional knowledge claims go unchallenged, or politics and knowledge are neatly separated also seems out of the question. In such an environment there appears to be a great need for a reconfiguration of truth that, while perhaps not solving the modernism/postmodernism impasse, allows for the generation of new languages, models of explanation, and methodologies.

Such a concerted effort to reconfigure truth can be found within the discipline of sociology. As the debates between modernists and postmodernists have intensified over the last twenty years, the discipline of sociology has, at least until recently, remained largely on the sidelines.[2] Even the sociology of knowledge, a branch of sociology that, since its inception, has been concerned with the relationship among power, context, and knowledge, has remained virtually silent about both the implications of postmodern theory and the debate between modernists and postmodernists. For postmodernists, sociology's silence is perhaps to be expected. As a child of modern positivism, the discipline cannot bring itself to critique its own scientistic foundations, nor is it possible for the sociological perspective to escape the reflexivist irony or "Mannheimian Paradox" that often results from the social realism implicit in most sociological analyses of knowledge and truth. Sociologists and sociologists of knowledge, like their modernist counterparts in philosophy and the natural sciences, all want, in one way or another, to maintain an epistemically privileged position.

Despite sociology's relative silence, in recent years it has produced a number of unique accounts of knowledge that have the potential of competing with both the scientific realism often advocated by modernists and the deconstructionist and textualist view of knowledge found in postmodernism. It has also produced accounts of knowledge that significantly redefine and redirect its century-old doctrine of social realism. These new and innovative approaches to understanding knowledge are outgrowths of a branch of the sociology of knowledge known generally as science studies or, more recently, science and technology studies (STS). Postmodernism and STS share the common goal of putting an end to realist epistemology and accompanying correspondence theories of truth. They both reject the traditional philosophical quest to find an epistemologically privileged position from which to mirror reality and the modernist argument that some inherent quality of the mind provides the prerequisite for the discovery of truth. As we shall see however, the two approaches are greatly divided over what the end of realist epistemology and the modernism/postmodernism debate mean for contemporary theory. They are also divided over the options that are now available to those concerned with assessing the validity of knowledge.

In this book I am concerned both with outlining the various options now open for social and cultural theorists of knowledge and with identifying some of the contributions the newer sociology of knowledge found in STS can make in the development of different conceptualizations of knowledge, reality, and truth. I contend that the approaches to knowledge found in STS have implications far beyond the circle of historians, philosophers, and sociologists to which they have been relegated. In this work, I consider the contributions and limitations of a number of classical and contemporary approaches in social theory and STS that fall under the general label of the sociology of knowledge. The specific focus of this book, however, is on the "amodern" approach to knowledge found in what is usually called actor-network theory (see Latour 1987, 1988a, 1993; Callon 1986, 1987). The intent of this work is not simply to provide an intellectual history of the origins and developments of actor-network theory, but to illustrate how this approach fits into the history of the sociology of knowledge and how it can contribute to the larger debate occurring in the humanities and social sciences on the possibility of obtaining objective, representational knowledge. In general terms, I wish to show how actor-network theory can be used to avoid the epistemic impasse of the current modernism versus postmodernism debate and to develop a more general reconfiguration of reality, knowledge, and truth. More specifically, I want to explore how an actor-network approach to knowledge may be particularly helpful in identifying the philosophical oppositions under which modern epistemology, postmodern antiepistemology, and even the social realism of the traditional sociology of knowledge often operate.

It is important to point out that I am not principally concerned in this work with the long-standing philosophical debate between realists and antirealists. That book has been written. In fact, it has been written many times over. In this regard, this work is not one of philosophy but one of sociology. In making this point I am not attempting to draw a rigorous distinction between the two approaches, as most nineteenth-century social theorists felt inclined to do, but rather to limit and in some ways redirect the current debate over knowledge. While the work of philosophers is used to illustrate modern and postmodern positions, I am not concerned here with pointing out the intricate substantive or historical details of realist or antirealist positions. Instead, the ideas of philosophers and epistemologists are used as a means for setting up the discussion of the sociology of knowledge and actor-network theory.

Overview of the Book

In scanning the various discussions currently taking place in academia over the status of knowledge, truth, and reality, it is possible to identify three common ontological perspectives and epistemological options. First among these is the perspective of scientific realism. In simplified terms, this perspective combines the ideas found in empiricist and rationalist philosophies of knowledge. Scientific realists maintain the existence of an absolute reality beyond consciousness and interpretative alteration. This absolute reality is translatable and, consequently, knowable if the proper scientific methods are utilized to recover it.[3] Second is the multifaceted perspective of postmodernism. This view builds on the arguments of earlier idealist and nominalist philosophies of knowledge.

Postmodernists deny the possibilities of ever grounding knowledge in some absolutist or naturalistic view of reality or in any unbending methodological procedures. They contend that all knowledge, regardless of its ontological status or methodological origins, is forever mediated by language and interpretation. This means that knowledge always and everywhere bears the mark of hermeneutical intervention—it is forever a product of the power to nominate and of rhetoric. The third perspective found in contemporary discussions of knowledge, truth, and reality is social realism. This position's origins can be found in earlier materialist philosophies of knowledge. Social realism shares some similarities with postmodernism, particularly with regard to rejecting naturalistic explanations of phenomena. Social realists, however, replace postmodernism's emphasis on language and signification with a sociological emphasis on society and social organization. Social realists maintain that knowledge, truth, and reality are in actuality the products of social forces, such as social organization, power relations, social interaction, and social hierarchies.

It is my primary goal in this work to add another perspective to the ones described above. More specifically, I wish to set up a four-way discussion among the above three positions: scientific realism, postmodernism, and social realism, with the actor-network theory found in the newer sociology of knowledge—a position that is critical of all three "isms". To do this, I examine in chapter 1 the framework and divisions found in the modern episteme, as well as how these divisions made possible the development and growth of the philosophy of scientific realism. The purpose of this chapter is not so much to provide a comprehensive history of modern epistemology or scientific realism, but to examine some of the boundary conventions and divisions that were constructed to separate science from nonscience and to make science a privileged way of knowing. I contend that these boundaries create the opportunity and motivation for the formation of postmodern deconstructionism in the late twentieth century.

In chapter 2, I investigate some of the manifestations of postmodernism's approaches to knowledge and its attempts to deconstruct the boundaries and divisions of the modern episteme. Specifically, I focus on the writings of Friedrich Nietzsche, Jacques Derrida, and Jean Baudrillard. I look at how their writings are partially responsible for three of the key moves of postmodern thought: the symmetricalizing of all knowledge claims, the call for a radically reflexive account of knowledge, and the subsequent transformation of all truth claims into signification, texts, and rhetorical forms.

In chapter 3, I explore how the general undermining of modern, realist epistemology found in postmodern theory has been utilized to deconstruct the foundationalist assumptions of modern science, as well as its more general claim to represent reality accurately. I focus on various recent attempts to treat science as a form of rhetoric, narrative, or text, as well as the efforts of feminist critics of science to examine the engendered origins and implications of scientific rationality and modern epistemology. Finally, I look at the recent response of scientists to the rhetorical and feminist treatments of science and how their response only intensifies the unresolvable epistemic impasse between modernists and postmodernists.

In chapter 4, I move the discussion toward the doctrine of social realism and traditional sociological explanations of knowledge. Here, I examine some of the theoretical roots of social realism and the sociology of knowledge. I argue that it is in the writings of Emile Durkheim that the boundary divisions of the modern episteme were most forcefully introduced into sociological analyses of knowledge. Essentially, I assert that Durkheim's writings provided social accounts of knowledge with three basic epistemic trajectories and theoretical impulses. One impulse sought to forge the doctrine of social realism by making sociological knowledge a special way of knowing. This was to be accomplished by applying scientific methods to the study of society. The second impulse attempted to use the doctrine of social realism to examine the relationship between social morphology and social representations, including knowledge, classification systems, and religion. This impulse eventually led to a third, more reflexivistic impulse. This impulse called for the examination all forms of knowledge, including science and social realism themselves. It is at this third, reflexivistic point where the sociology of knowledge turns on itself and becomes highly paradoxical.

Chapter 5 explores some of the various twentieth-century attempts in the sociology of knowledge to come to terms with the legacies left by Durkheim, specifically his problematic relationship between social representations of knowledge and sociological rerepresentations of that knowledge. I examine several strategies used in the twentieth-century sociology of knowledge, such as those found in Karl Mannheim, Robert Merton, ethnomethodology, the sociology of scientific knowledge, social constructionism, and the more recent reflexivity project. I will show how these strategies, while often empirically productive, fail in the end to overcome the legacies of Durkheim or to answer the important questions about knowledge now posed by postmodern theory.

In chapter 6, I begin the exposition of actor-network theory's empirical model of knowledge formation. Here, I consider the basic actor-network model of the production of facts advocated by Bruno Latour. I also consider some of his rules of method and how these rules allow for a different explanation of the modern division between science and nonscience. Finally, I explore some of the implications of Latour's "railroad" or "cable-TV" model of reality making.

Chapter 7 surveys actor-network theory's assessment of modernity and modern ways of knowing. In this chapter, I examine Latour's critique of what he refers to as the modern constitution of truth and how, within his view, the proliferation of hybrids has led to its demise. I also consider his proposal for an amodern or nonmodern constitution of truth and how he seeks to avoid some of the problems found in modern and postmodern conceptualizations of knowledge, truth, and reality.

In the final chapter, I examine some of the possibilities opened up by actor-network theory for reconfiguring the positions found in scientific realism, postmodern textual relativism, and social realism. I also explore the impact of the actor-network perspective on the modernism versus postmodernism impasse. I argue that the views of knowledge found in postmodern theory and the social realism of sociology of knowledge both suffer from the "spell of representation." They therefore maintain a type of epistemological vigilance implemented in the modern way of knowing. Finally, I suggest ways in which actor-network theory enables the sociology of knowledge and other accounts of knowledge to bypass the plaguing issue of reflexivity that has captivated disciplines for well over a century.

Notes

1. Without doubt, the debates between modernists and postmodernists include more than discussions of truth, reality, and knowledge. The debates also involve related and divergent issues concerning the possibilities of emancipatory politics (Habermas 1981, 1987; Sloterdijk 1987; Seidman 1991), architecture and art design (Jencks 1987; Portoghesis 1983; Mahsun 1987), literary form (Hutcheon 1989), the impact of late capitalism (Jameson 1991; Lash and Urry 1987), and a countless variety of other issues and topics.

2. While the modernism/postmodernism debates have been incorporated into some areas within sociology, particularly feminist theory, social theory, and cultural studies, generally speaking, the discipline as a whole has been somewhat slow in coming to terms with the so-called postmodern turn and its epistemological revelations and implications. For instance, it was only in the late 1980s that articles began appearing in sociology journals addressing issues related to postmodernism and postmodernity (see Featherstone 1988). This awareness in sociology came years after the issues associated with postmodernism had been raised in the fields of philosophy and literary criticism.

3. The idea that nature selects those statements that correspond to reality is not just the position of many modern philosophers of science, but also represents the views of many scientists actually practicing science. For a discussion of this see Pickering's *Constructing Quarks* (1984) and Peacocke's *Intimations of Reality* (1984).

Chapter 1

The Boundary Conventions of the Modern Episteme

Definitions and descriptions of modernity generally focus on large-scale changes in society and distinct cultural and historical characteristics.[1] Modernity, however, also involves certain unique modes of constituting and organizing reality and truth (see Foucault 1970, 1972; Law 1994). In this sense, modernity can be seen as imposing particular ontological, epistemological, and methodological prescriptions, divisions, and delimitations on reality and truth. Foucault (1972) labeled these ways of organizing and limiting reality and truth as "epistemes." He defined epistemes as "the total set of relations that unite, at a given period, the discursive practices that give rise to epistemological figures, sciences and possible formalized systems..." (Foucault 1972, 191-192). In Foucault's view, epistemes could be thought of as truth regimes or political systems of knowledge seeking. Seen from this Foucaultian perspective, truth is not a neutral result of knowledge seeking, but an activity that always "permits and assures the exercise of power" (Foucault 1979, 62). Epistemes provide a political means of orienting truth-finding practices and knowledge production.

Epistemes and epistemic divisions, however, are not just entities exclusively concerned with the political organization of knowledge. They are also systems that through the process of organizing knowledge also organize people. In this process, epistemes impose moral and symbolic boundaries among individuals and social groups (see Shapin 1994). Like all moral and symbolic boundaries, these epistemic boundaries draw divisions between correct and incorrect ways of thinking, right and wrong procedures and practices, and, ultimately, good and bad people. They serve to impose and enforce moral codes of conduct, cognitive standards, and economies of judgment. Seen from this perspective, the modern episteme can be conceptualized as a means of setting up certain unique rules for the production of knowledge, as well as exclusionary mechanisms and regulatory laws for the enforcement of those rules.[2] It serves to permit and prohibit different kinds of cognitive moves and styles.

In this first chapter, I want to examine some of the central boundary conventions and divisions of the modern episteme. I start with the basic assumption that much of social life, including epistemic frameworks and knowledge production, is the result of "boundary work" (see Gieryn 1983). Boundary work is viewed here as an intricate process of negotiation, conflict, and compromise where limitations and opportunities are constructed, implemented, and enforced. As Mary Douglas (1985, 5) has described the process: "All the arguments taking place in families, churches and sports clubs are about whether the institution shall draw its group boundary closer, or relax it, apply its rules more strictly, create more rules or relax them all."

Thus, epistemological boundaries can be thought of as social or associational mechanisms through which reality is ordered and maintained. In turn, these boundaries turn into "forms of social organization and...a means of regulating social interaction" within knowledge-producing communities (Shapin and Schaffer 1985, 14). They serve as both political and moral means through which intellectual authority and property are acquired, enlarged, and defended.

In this chapter, I focus primarily on the general modernistic boundary between science and nonscience, as well as the corollary divisions that support it. To do this, I will examine some selected seventeenth- through nineteenth-century writings of advocates of science, social science, and natural philosophy. Such a selective inclusion is not meant to suggest that the writers examined here are solely responsible for the origin of the modern way of knowing, that they wrote the definitive word on the modern episteme, or that they were unopposed in their efforts. Rather, the modern episteme is seen here as a product of the on-going boundary and craft work performed over the last four or five centuries. In fact, it could be argued that much of this modern episteme can be seen as the post hoc construction of philosophers and historians of science in the twentieth century and that the divisions of the modern episteme are not to be seen here as foundational or nonreversible. The boundaries of the modern episteme, like all boundaries, are seen as "ambiguous, flexible, historically changing, contextually variable, internally inconsistent and sometimes disputed" (Gieryn 1983, 792). Boundaries are always fragile and contingent accomplishments. They are always in danger of being toppled and replaced by the boundaries and divisions of other groups. As such, they require constant implementation and reinforcement in order to be maintained and enlarged. In addition, the divisions are viewed here as carefully crafted human products, not closer reflections of reality or the results of unique ontological conditions. It is part of the general realization "that it is ourselves and not reality that is responsible for what we know" (Shapin and Schaffer 1985, 344). In fact, it is the malleability of these conventions, as I argue in the next chapter, that makes the postmodern attempt to reconfigure these boundary conventions possible and even likely.

In order to explore the ontological and epistemic structure of the modern epistemic division between science and nonscience, I focus in this chapter on three central boundary conventions of the modern way of knowing: the overlapping and interrelated divisions between belief and truth, rhetoric and reality, and literary and scientific knowledge.

Belief versus Truth

No boundary convention is more important to the modern way of knowing than the one between belief and truth. The modern episteme sets up firm distinctions between legitimate and illegitimate types of knowledge claims, including a categorization of the claims that constitute knowledge and the types of people who can make those claims. Legitimate knowledge claims are seen as those that adhere to a certain cognitive style, use the scientific method, and are open to critical evaluation and reformulation. Illegitimate knowledge claims are seen as those that are derived from undisciplined lay accounts of phenomena and haphazard techniques and are simply accepted at face value. The lexicon produced by the modern episteme is filled with a multitude of terms used to describe this separation of truth from belief. Ideas and explanations that are produced within the modern episteme are said to be "accurate," "rational," "modern," or "scientific." Conversely, claims that are outside of this epistemic framework are said to be "misinformed, "irrational," "primitive," or "mythical." Knowledge produced within the modern episteme is viewed as simply of a different type and higher order than the knowledge produced from other ways of knowing. In fact, those "other ways of knowing" do not really constitute knowledge at all, but are only beliefs.

Historically, advocates of the modern episteme have gone to great lengths to separate the believers from the knowers or the ordinary from the specialized. One avenue sought to demarcate belief from knowledge by focusing on the cognitive characteristics or qualities of the mind. From the position of the modern episteme, the primitive or lay mind is thought to be a product of cultural traditions and superstitions, while the modern mind is seen as being unrestrained by cultural traditions and, hence, potentially pure. Under the influence of cultural or religious traditions, the primitive or lay mind is thought to be unreflective and mechanical, while the modern or scientific mind is said to be reflective and methodical. The cognitive characteristics of the primitive or lay mind lead to the production of myth, religion, folklore, and lies to explain the world, while the modern mind produces the analytics and experimentalism of philosophy and science.

Evidence for such a view of the cognitive distinction between various types of people can be found throughout seventeenth-century philosophy and science. For instance, John Locke (1975, 715) declared that "earthy minds, like mud walls, resist the strongest batteries." Even when ordinary minds confront "the force of a clear argument" they "nevertheless stand firm, keep out the enemy truth that would captivate or disturb them" (Locke 1975, 715). Likewise, in Francis Bacon's view, most minds are trapped in the superstitions of tradition. "Our spirits are included in the caves of our own bodies; so that they must needs be filled with infinite errors and false appearances" (Bacon 1961a, 433). He argued that "it is not the lie that passeth through the mind, but the lie that sinketh in and settleth in it" (Bacon 1961b, 378). He compared the mind to "an enchanted glass, full of superstition and imposture" (Bacon 1961a, 431). The mind "distorts and discolours the nature of things by mingling its own nature with it" (Bacon 1961a, 54). The mind was capable of overcoming lie and superstition only if it could "lay aside received opinions and notions…and refrain the mind for a time from the highest generalizations" (Bacon 1961a, 115). Bacon (1961a, 115) believed that if this was accomplished then individuals "would be able by the native and genuine force of the mind, without any other art, to fall into my form of interpretation." In other words, if the mind were free from the cultural superstitions that limited and confined its scope, then it could reach the proper rational interpretation of phenomena based on the empirical evidence Bacon advocated.

One of the most central strategies found in the modern episteme to separate belief from knowledge was to distinguish the type of knowledge derived from religion from that obtained from science and philosophy.[3] Most advocates of early science or natural philosophy were firm believers in religious and theological principles; however, many were determined to either separate theological truths from experimentally derived ones or to adjust religion to conform with science (see Westfall 1958). From the perspective of advocates of early modern science, "to seek philosophical truth in the Bible would be to confuse the boundaries of the two forms of truth" (Westfall 1958, 163). The ontological status of different things (e.g., spiritual versus natural) required different types of methods, proofs, and evidence. In the *Principles and Duties of Natural Religion*, John Wilkins (1675, 25) argued that "men ought to rest satisfied in the best evidence for it which that kind of things will bear and beyond which better could not be expected supposing it were true." God had divided the world into things that were knowable by humans and things that were knowable only by God. Physical reality was the bastion of human knowledge; however, knowledge of the soul or spiritual matters were the exclusive domain of God. In this scenario, humans had to be content to study those things they were, by the grace of God, capable of grasping. What was not spiritual in origin was open to the scrutiny and explanation of human observers.

Some scientists and philosophers, like Robert Boyle and Locke, set about the task of preserving both religious and scientific thinking. Boyle argued that religion was "too important to be accepted on trust" (Westfall 1958, 168). Consequently, religious claims should be opened to limited rational scrutiny. There were, however, still many elements of reality, or, in his terms, "privileged things" that were unknowable by humans. From his view, philosophical endeavors were always incomplete and in need of refinement. Belief, however, still required some types of evidence. Locke, too, was a proponent of a reasoned approach to religion. He believed that it was possible carefully to "lay down the Measures and Boundaries between Faith and Reason" (Locke 1975, 688). Locke (1975, 689) defined reason as "the discovery of the Certainty or Probability of such Propositions or Truth, which the Mind arrives at by Deductions made from such Ideas, which it got by the use of its natural Faculties." In contrast, he defined faith as "the Assent to any Proposition, not made out by the Deductions of Reason; but upon the Credit of the Proposer, as coming from GOD, in some extraordinary way of Communication" (Locke 1975, 689).

Others, such as John Tyndall, a nineteenth-century member of the Royal Society, took a more skeptical view of religion and the impact of religious thinking on knowledge. Tyndall declared that "foolishness is far too weak a word to apply to any attempt to force upon a scientific age the edicts of a Jewish lawgiver" (Tyndall 1893, 36). Tyndall was a firm defender of scientific versions of truth in the face of religious speculation and belief. From his view, religion was a useful aid for comforting people but an ineffective means for deciding issues of absolute and useful truth (see Gieryn 1983, 785). Science produced accurate and useful knowledge, while religion produced speculation and mythical explanations. In Tyndall's view, as long as religion was seen as a form of poetry "it is forever beautiful"; however, if seen as an absolute truth, it is "purely obstructive and hurtful" (Tyndall 1905, 224). In other words, as long as religion stayed within its place and did not cross the boundary between belief and reason, it was acceptable as a necessary mythical and psychological aid. If it violated its position, however, and began making pronouncements of certainty and absolute truth, it was a dangerous threat to the accumulation and advancement of knowledge.

According to many advocates of natural philosophy, another central distinction that separated belief from knowledge was the types of methods used to derive knowledge. Within this distinction, knowledge was said to be based on a rigorous methodology that was exact or exactable, while everyday knowledge relied only on hearsay and second-hand reports. Francis Bacon argued that natural philosophy must spend more time observing nature rather than relying on the verbal truth passed on from belief (see Slaughter 1982, 90). Such a view required a means for disciplining and controlling the mind. The mind must not be "left to take its own course, but guided at every step; and the business be done as if by machinery" (Bacon 1961a, 41). He also argued, "When a man tries all kinds of experiments without order or method, this is but groping in the dark; but when he uses some direction and order in experimenting, it is as if he were led by the hand; and this is what I mean by Learned Experience" (Bacon 1961a, 413). Methodology was seen as a means for imposing discipline on investigators and the investigating process. Traditional forms of knowledge had relied on hearsay and personal accounts to formulate knowledge. This type of undisciplined exploration of events and phenomena was unacceptable for those who wanted to know the true nature of objects. They needed and demanded discipline and procedure.

A final feature of the division between belief and truth involved the type of environment in which knowledge was produced. In this view, scientific knowledge was extraordinary because it was produced in a special place with special people and special equipment. The special place was the scientific laboratory where experiments were created and conducted. The laboratory allowed for the collective witnessing of experiments. It allowed for a claim "to be legitimized by a corporation of scientific practitioners rather than by an individual patron" (Biagioli 1993, 357). The special people were the gathering of "gentlemen" whose integrity, trustworthiness, and social standing could not be questioned. These honorable and trustworthy gentlemen were used to witness and verify the findings of experiments (see Rudwick 1985; Shapin 1994).[4] Unlike the lay knowledge of everyday life, which often relied on rumor from unscrupulous sources, the laboratory brought together a group of like-minded peers who could see for themselves the outcome of experiments and testify to experimental truthfulness. As in a court trial, witnessing was a crucial aspect of establishing experimental matters of fact and moral certainty (Shapin and Schaffer 1985, 56; Shapiro 1983). Thus, truth was obtainable to those who practiced an appropriate "epistemological decorum" (see Shapin 1994, 193). It required an appropriate style of conduct that only upstanding members of the community could adhere to. In this sense, the experiment itself was "more than a method of discovery; it [was] an ordeal, a test of its subject's true nature" (Briggs 1989, 3). As Bacon (1961a, 26) put it, "the office of the sense shall be only to judge of the experiment, and that the experiment itself shall judge of the thing."

The special equipment of this extraordinary knowledge was the instruments of the laboratory. As with practicing a rigorous methodology, scientific instruments were also seen as means for correcting and disciplining the senses (Shapin and Schaffer 1985, 37). The senses were often capable of being misled by visual distortions or other sensory apparitions. By using specialized equipment, the senses were capable of being better organized and regimented in order to examine and record natural events accurately.

By the eighteenth century the boundary between belief and knowledge had become so complete that the Marquis de Condorcet was able to predict that "the time will come when the sun will shine only of free men who know no master but their reason." (Condorcet quoted in Jones 1955, 2). Such optimism led Condorcet (1955, 4) to argue that it was now possible to "show by appeal to reason and fact that nature has set no term to the perfection of human faculties; that the perfectablity of man is truly indefinite." The sciences of nature and the newly emerging sciences of society would pave the way for the enlightenment and infinite perfection of all of humanity. Herein, the first boundaries of the modern episteme begin to be erected.[5]

Rhetoric versus Reality

Another key distinction found in the modern episteme's division between science and nonscience was one between rhetoric and reality. In Western philosophy, rhetoric and its corollaries poetry and literature have been treated as the antithesis and enemy of reality and truth since the Platonic dialogues and Plato's attack on sophism. For Plato, rhetoric was incapable of providing true insight or real knowledge: it only "urges on us convictions or opinions" (Ijsseling 1976, 7). The Platonic rejection of rhetoric took on not only an epistemological imperative but also a moral one. In Plato's view, rhetoric not only perpetuated opinions but was also an impediment to the moral enlightenment of humanity and its approximation of the good form. For instance, Plato (1966, 340) felt that poetry "has no serious claim to be valued as an apprehension of truth." Like a lover who shuns "a passion which he finds is doing him no good," society should ban poetry until poets could show in lyrical form how it is beneficial (Plato 1966, 340).

Despite Plato's attempts to subdue rhetoric via the logical rigor of the dialogues and, later, Aristotle's efforts to break knowledge into the distinct and ranked domains of theoria, praxis, and poesis, rhetoric continued to be emphasized and taught throughout the subsequent centuries. The assault against rhetoric again intensified, however, with the rise of the New Science and its "experimental epistemic culture" in the seventeenth century (Fuller 1993, 110). Here, rhetoric was again condemned as the enemy of knowledge, reality, and truth (see Dentith 1990, 2). This time, however, unlike in Platonic philosophy, the criticism of rhetoric was made not in defense of logic and philosophical authority, but in the promotion of the empirical methods of the newly emerging experimental sciences.

Despite philosophical disagreement between various advocates of science and philosophy, most were in agreement that rhetoric was something that had to be limited and contained in order for knowledge to develop. Accounts of New Science's attack on rhetoric were widespread in the seventeenth century. Descartes (1952, 31–32) argued that science needed only the strength of its arguments to persuade, not the manipulation of rhetoric. In his view, one of the principle causes of errors in thinking was the relationship between words and real things (see Descartes 1988, 187–188). He maintained that for knowledge to grow it was important to be able to "distinguish those of our concepts which are clear and distinct from those which are obscure and confused" (Descartes 1988, 188).

Likewise, Bacon attacked the role of rhetoric in reproducing errors and superstitions. He argued that "logic handleth reason exact and in truth, and rhetoric handleth it as it is planted in popular opinions and manners" (Bacon 1961c, 411). Bacon condemned the play of language, or what he termed the "Idols of the Marketplace" and the theatrics of rhetoric, or the "Idols of the Theatre" (Bacon 1960, 56–60). The idols of the theater mislead people by confusing the truth of the matter with fancy phrases and theatrical props. Bacon wrote (1961a, 63): "And in the play of this philosophical theatre you may observe the same thing which is found in the theatre of the poets, that stories invented for the stage are more compact and elegant, and more as one would wish them to be, than true stores out of history." The various idols "impose themselves upon the fancies of the learned so long as men do not embark upon the road of true experiment" (Briggs 1989, 196). In Bacon's view, words block and resist the advancement of knowledge (see Bacon 1961a, 434). Words "follow those lines of division which are most obvious to the vulgar understanding" (Bacon 1961a, 433). For Bacon, the problem was not just that rhetoric confused reason, but that it overwhelmed and subdued the imagination. He declared, "away with antiquities, and citations or testimonies from authors and also with disputes and controversies and differing opinions—everything, in short, philological" (Bacon 1960, 274). Since words distorted understanding, Bacon urged scientists to report experiments "briefly and concisely, so that they may be nothing less than words" (Bacon 1960, 274). Like Bacon, Isaac Newton was also very concerned with the impact of inaccurate language on scientific endeavors. He argued for the creations of a "general language" (Newton quoted in Slaughter 1982, 153). This general language would be drawn from "the natures of things themselves" and not the distortions and misrepresentations of existing languages.

Other evidence of the disdain for rhetoric can be found in the late seventeenth-century proceedings of the Royal Society. One of the society's explicit programmatic goals was to "separate the knowledge of Nature, from the colours of Rhetoric, the devices of Fancy, or the delightful deceit of Fables" (Sprat 1958, 62). The Royal Society required that all reports be given "without any Prefaces, Apologies or Rhetorical Flourishes" (quoted in Jones 1971, 461). In the view of the Royal Society, no "vanity of fine speaking," trick of metaphor," or "volubility of Tongue" was needed to persuade others of the importance of one's findings (Sprat 1958, 112). The witnessing of experiments or the "virtual witnessing" in scientific articles meant that experimental facts could stand on their own (Shapin and Schaffer 1985, 60; Bazerman 1988, 74).

In the eighteenth century, the task of cleansing the empirical sciences from the contamination of rhetoric and language was undertaken by Enlightenment figures such as Etienne de Condillac and the Marquis de Condorcet. Condillac (1980, 16) described his intellectual mission as "unraveling the chaos into which the abuses and vices of language have plunged the moral and metaphysical sciences." He argued that "reasoning is perfected only to the extent that languages are themselves perfected" (Condillac 1980, 303). Condillac declared that the solution to the distortions of language were to be found in the "well-made" language of algebra. In his words:

> Algebra is very striking proof that the progress of the sciences depends solely upon the progress of their languages; and that well-made languages alone could give to analysis the degree of simplicity and precision of which it is capable in each area of our studies. (Condillac 1980, 305)

Condillac contended that the perfection of language would allow for all realms of inquiry to be opened to the advancement of reason. Likewise, Condorcet (1955, 185) argued that

> As the number of facts increase, the human mind learns to classify them and to subsume them under more general facts, and, at the same time, the instruments and methods employed in their observation and their exact measurement acquire a new precision the language that fixes and determines ideas will acquire greater breadth and precision.

Part of the rejection of rhetoric found in early modern science and philosophy derived from the scientific view of the relationship between objects and representation. From this perspective the world was "a fixed object of analysis quite separate from the forms of discourse by which men speak of it and by which they represent their thoughts" (Reiss 1982, 24). Advocates of early modern science were captivated by the task of establishing a direct relationship between objects and representation (Shapiro 1983, 227). Objects could be accurately represented if, and only if, proper methodological procedure was followed and evidence was available for all to observe readily (see Rudwick 1985; Ophir and Shapin 1991, 11; Shapin 1994). Furthermore, while representations of nature might occasionally miss their mark, usually as the result of misapplied methodology or human error, they could, with the further refinement of technique and the evolution of knowledge, come closer and closer to the real. Essentially, early modern science sought to use experimental results to construct a universal, descriptive metalanguage that was capable of avoiding the imprecision of ordinary speech and the misleading character of rhetoric (see Shapiro 1983, 227; Slaughter 1982).

In the view of early scientists, objects of nature could be described in terms of their mathematical dimensions, classification type, or chemical composition, rather than by their common cultural use. Avoiding the use of rhetoric in writing scientific articles thus became a means for scientists to show that they were "aware of the working of the Baconian idols and were taking measures to mitigate their corrupting effects on knowledge claims" (Shapin and Schaffer 1985, 68). It meant that scientists were taking all the necessary precautions to ensure that persuasion and argumentation were based on the weight of the evidence alone and not the manipulations and distortions of rhetoric.

Literary versus Scientific Knowledge

The New Scientists' rejection of rhetoric can be seen not only as an attack on the distorting effects of rhetoric on scientific inquiry, but also as an assault on one on their chief knowledge-production rivals—literary humanists. In this sense, the division between science and literature and scientists and humanists can also be seen as one of the other key demarcations found in the modern episteme. Advocates of New Science often attacked those "who copy books in lieu of observing nature" (Jones 1961, 20). In *De Magnete*, written at the turn of the seventeenth century, William Gilbert (1958, 47) argued:

> Men are deplorably ignorant with respect to natural things, and modern philosophers, as though dreaming in darkness, must be aroused and taught the use of things, the dealing with things; they must be made to quit the sort of learning that comes only from books, and that rests only on vain arguments from probability and upon conjectures.

Gilbert (1958, 5) criticized literary and philosophical humanists who were only "acquainted with books" rather "than being practical in the research of objects in nature." Galileo also made similar claims about the knowledge limitations of the humanists. In a letter to Johannes Kepler, he wrote, "this sort of man thinks that philosophy is a sort of book like the Aeneid and Odyssey, and that truth is to be found not in the world or in nature but in the collation of texts" (quoted in Grafton 1991, 2).

Bacon, too, argued that the knowledge produced by humanists was of no practical use for experimentalists. He argued that "those foolish and apish images of worlds which the fancies of men have created in philosophical systems, must be utterly scattered to the winds" (Bacon 1961a, 110). He felt that there were writers "everywhere parading similitudes and sympathies of things that have no reality" (Bacon 1961a, 167). Bacon combined all previous knowledge forms into one category: "the making of knowledge into an institution, a matrix, a text" (Whitney 1986, 131). These old types of knowledge produced primarily by "Speech and Argument" were deficient ways of accumulating knowledge and, therefore, to be avoided (Bacon 1961c, 384).

Attacks on literary knowledge and those who produced it were certainly not limited to the seventeenth century. In the early 1960s the biologist Bentley Glass (1960, 155) warned the humanities and social sciences that they "must be permeated with the knowledge and spirit of science if they are to be more than relics of a buried age." From this view, science was the model of all knowledge seeking. Those who ignored its power and persuasiveness would inevitably fall by the wayside as scientific knowledge and methodology continually improved and expanded into new domains.

While the rhetoric directed at literary humanists was often harsh, proponents of the modern episteme should not be seen as instigating an unmediated attack on humanists and their knowledge products. In actuality, most of the advocates of the New Science, as well as most of the Enlightenment philosophers, were strong proponents of the type of liberal learning acquired through poetry, literature, and other nonscientific forms of knowledge (Grafton 1991). What occurred here, as with religion, was an attempt to limit the scope and range of the types of claims that humanists could make. Essentially, advocates of the New Science wanted to forge an intellectual and political division of labor that would compartmentalize knowledge into distinct spheres and assign each party a proper object and domain of expertise. Within this system, "objects and subjects would not and could not be treated as part of the same philosophical enterprise" (Shapin and Schaffer 1985, 337). Humanist knowledge and its interpretative and speculative methodology were to be limited to human affairs, such as the analysis of politics and culture. They were not directly applicable to the study of natural phenomena or for providing rules of scientific methodology. That sphere was to be the domain of scientists and their experimental approach to knowledge.[6]

Conclusion: The Modern Episteme and the Development of Scientific Realism

The modern episteme, as it has been called here, and its exclusion of the fantastic from the real "makes modern truth possible" (Aichele 1994, 323). It also allows for and vindicates the development of the doctrine of scientific realism. Scientific realism, like its more encompassing cousin philosophical realism, has a number of different versions and manifestations (see Vision 1988). Its most basic proposition, however, is that science, and only science, is capable of providing accurate representations of reality. Scientific representations are, then, "factually true generalizations about the actual behavior of real physical objects existing in nature" (Rescher 1987, 4). Scientific realism further argues that objects exist independently of human cognition and that concepts and theoretical generalizations, if properly formulated, are capable of conforming to the reality of those objects. In stronger terms, statements about reality are always "true or false in virtue of a reality existing independently of us" (Dummett 1978, 146). Scientific realism draws a firm philosophical distinction between the world as it is perceived and the world as it actually is. Truth is defined as those statements that successfully cross the ontological boundary between subject and object and successfully capture the inherent nature of that independent reality.

Scientific realists also contend that scientific methodology serves as the only reliable strategy for unveiling reality. Furthermore, as the knowledge generated by that methodology accumulates, science discovers better and better explanations of reality and converges on the truth (see Kitcher 1993, 127; Leplin 1984a; van Frassen 1980; Boyd 1984).[7] Because the scientific method imposes a careful and tireless research strategy, it is viewed as the only genuinely reliable means for uncovering the real. Only the method and institution of science are capable of setting up the experimental and organizational circumstances necessary for discovering and reporting the inherent nature of objects. The various branches of scientific realism are united in the belief that scientific representations can, under the special circumstances of scientific experimentation, mirror, or at least approximate, reality and report truth (see Rorty 1979; Hacking 1983, 5).

This view does not deny that alternative and competing theories are possible, only that it is nature and not the scientists themselves who ultimately decides which theory becomes accepted truth. Repeated experimentation and research will eventually reveal which theory holds and which one does not. The merits of a claim or theoretical position are to be judged exclusively by its congruency with an independent, yet knowable, reality.

For advocates of scientific realism, theirs is the only view that accounts for the important discoveries of science and that "does not make the success of science a miracle" (Putnam 1975, 73; see Leplin 1984b, 1). Here, the ability of science to provide workable manipulations of objects and accurate predictions of nature is used as proof of its success and epistemological supremacy. In other words, advocates of scientific realism argue that science is both successful and truthful because it speaks the language of nature. Science's representations hold because they really do reflect reality.

In the twentieth century, scientific realism has served as a symbolic means for organizing and unifying the set of ideas and practices found in the modern episteme. As I have tried to illustrate in this chapter, however, the modern episteme is not just a group of ideas about the nature of reality and truth. It also exemplifies a series of values and moral convictions, not only about truth and reality, but also about the social organization and political boundaries of knowledge production. The modern episteme contains provisions about and divisions between good and bad forms of life: a division C.P. Snow (1959) referred to as the "two cultures." By the mid-twentieth century it appeared to Snow and others that the division between scientific and literary fields was so great that "even on the level of emotion, they can't find much common ground" (Snow 1959, 5). Generally, these values and convictions of the modern episteme have not been seen as such, but simply as improved conceptualizations of the way reality actually works and as refinements of the means through which it can be captured. From this view, humanists have their place, but it is not at the seat of knowledge. They may enhance the appreciation of everyday life, but they are not capable of contributing to the advancement of knowledge and truth.

A central feature that runs through all the boundaries of the modern episteme, as well as the philosophy of scientific realism, is the great divide between interpretative and scientific knowledge and knowledge-producing fields. Historically, this division between fact and fiction and science and literature has been depicted as if it were merely an epistemological division between proponents of a scientific realism and advocates of a humanistic relativism. As a result, the divisions among the subject matter, methodologies, and knowledge styles of these fields and camps have become so acceptable in the modern academy as to appear as if they derive from an ontological division between nature and culture. Such a view, however, clouds the historical, social, and political struggles that have shaped and constructed this "natural attitude" and political boundary.

Since the seventeenth century, scientific and literary fields and camps have followed separate theoretical and methodological paths. This has been caused, at least in part, by a tacit agreement that divides knowledge into distinct ontological spheres and grants each group its own intellectual property rights. Since this division of intellectual labor was first developed and enforced by early natural philosophers and experimentalists, science has become the dominant institution of knowledge production and has been able to export its knowledge style and to colonize other fields. Science has even become so powerful as to be able to define the rules, rituals, and symbolic manifestations through which all serious speech is spoken and all important knowledge claims are to be examined and assessed. It "is near to being *the* source of cognitive authority: anyone who would be widely believed and trusted as an interpreter of nature needs a license from the scientific community" (Barnes and Edge 1982, 2). Those claims that do not adhere to the scientific version of truth, such as those in the humanities, art, or lay knowledge, are dismissed as speculative, amateur, subjective, pseudo, or soft and are relegated to the margins of knowledge production (see Collins and Pinch 1982). Science, not the literary fields, have become the modern means to access "worldly power" (Fuller 1993, 109). In the twentieth century, the modern episteme has allowed science to become "the measure of all things" (Sellars 1967, 173).

Notes

1. For example, such descriptions can be found in Karl Marx's explanation of the rise of industrial capitalism, Max Weber's account of the rationalization process, and Emile Durkheim's discussion of mechanical and organic solidarity and anomie.

2. These rules and mechanism have been labeled different things by different theorists; e.g., "paradigms" (Kuhn 1962), "epistemic cultures" (Fuller 1993), or "constitutions" (Latour 1993).

3. In some ways this parallels the separation of everyday life from heavenly affairs, a process described by Weber in *The Protestant Ethic and the Spirit of Capitalism* (1958;

also see Merton 1938).

4. Seth Ward once argued that in examining the merits of a truth claim, it was always advisable to scrutinize "the qualities of the relators". (Seth Ward quoted in Shapin 1994, 229).

5. The separation of belief and knowledge should be seen as a boundary convention that has been fought out and enforced over the course of the last few centuries. For instance, the debates between evolutionists and creationists or between theologists and genetic engineers are often depicted as struggles between the forces of reason and rationality and the defenders of belief and superstition. Even in cases where belief and knowledge obviously intersect, such as in the euthanasia debate, there are often struggles to control the types of arguments that those only possessing belief are able to put forward. These disputes are contemporary testimonies to the ongoing struggle to maintain the constructed distinctions between belief and knowledge that began in the seventeenth century.

6. An interesting outcome of the seventeenth-century division between literary and scientific knowledge can be found in the late nineteenth- and twentieth-century debates within the social sciences on methodology. As we shall examine in more detail in chapter 5, the social sciences prove to be one of the most interesting cases of the ongoing division between literary and scientific knowledge. For the most part, social scientists have never completely decided whether their disciplines are interpretative ones, modeled after the humanities, or experimental ones, modeled after the natural sciences. Since the methodenstreit in German social theory, some advocates of social science have pursued issues of interpretation and constructed meaning (e.g., cultural sociology and anthropology), while others attempt to match scientific expectations through a mathematical and experimental reconfiguration of objects (e.g., quantitative sociology and archeology; see Turner and Turner 1990).

7. In the twentieth century, two of the most influential versions and corrections of scientific realism have been Carnap's (1967) inductive and empirically based defense of science and Popper's (1959) more deductive description of falsification and the limitations of scientific explanations.

Chapter 2

The Revenge of the Humanities: The Postmodern Textualization of Truth

If it is possible to conclude that the modern way of knowing was built on an epistemic distinction between science and nonscience and corollary divisions between belief and truth, rhetoric and reality, and literary and scientific knowledge, then much of what is now called postmodern theory, at least in the humanities and social sciences, can be seen as an attempt to undermine or blur those epistemic distinctions. This blurring has had the latent effect of reconfiguring the sociomoral boundaries between experimental science and lay and literary speculation. It therefore dismantles, or at least attempts to dismantle, the political boundaries that have long divided scientists and humanists and the moral boundaries that have separated their knowledge forms and products.

In this chapter, I want to explore how the blurring of epistemic and moral boundaries has been accomplished. To do this I will focus on three writers associated with the formation of postmodernism, Nietzsche, Derrida, and Baudrillard. While these three theorists certainly do not exhaust the complexities of postmodern theory or provide a comprehensive description of its multifaceted positions, I do believe that their writings provide an avenue into some of the key positions of postmodernism, including the fascination with the relationship between representation and reality and the general reassessment of the epistemic status of all knowledge claims. Such fascination with representation is, of course, quite old in the Occidental theoretical tradition (e.g., Greek Sophism and Skepticism). However, what differentiates postmodern writings on the relationship between representation and reality from their sophistic or skeptical predecessors is their scope and radical reflexivity. Postmodernists are not content with only attacking certain types of discourse, but also take aim at all foundational claims, including sociology, philosophy, literary criticism, science, and even the latent foundationalist assumptions of their own discourse (see Fish 1989). In postmodern theory no field of inquiry, methodology, or mode of interpretation is safe from the deconstruction and textualization of truth. Literally everything is open to the swirl of signification and, consequently, the unending play of language.

Nietzsche on Truth and Lie

One of the most forceful attempts to undermine the boundary conventions of the modern episteme can be found in the late nineteenth-century writings of Friedrich Nietzsche. Nietzsche's primary strategy for subverting the modern episteme was to attack two of its most cherished components—the ideas of foundational truth and the rational subject. In this sense, his notorious proclamation of the death of God can be read not as just another late nineteenth-century statement on the secularization of Occidental culture, but also as a declaration of the end of all moral and epistemological foundations. God certainly was dead for Nietzsche, but so too was "everything that, in rapid succession, [had] tried to takes its place—e.g. the ideal, consciousness, reason, the certainty of progress, the happiness of the masses, culture, etc." (Blanchot 1986, 121). He predicted that the "history of the next two centuries" would be spent dealing with "the advent of nihilism" that the death of God and all foundational discourse had created (Nietzsche 1968, 3).

In contrast to modern realists, Nietzsche maintained that truth was not a fixed state that could be apprehended and spoken by pure reason, but a sociopolitical, rhetorical, and, hence, historically variable construction. In his words, truth was a "mobile army of metaphors, metonyms and anthropomorphisms—in short, a sum of human relations, which have been enhanced, transposed and embellished poetically and rhetorically" (Nietzsche 1968, 46–47). In this process, humans have forgotten "that the original perceptual metaphors are metaphors and take them for the things themselves" (Nietzsche 1989, 86). "Truth is therefore not something there, that might be found or discovered—but something that must be created and that gives name to a process." (Nietzsche 1968, 298). Like "coins which have lost their pictures," the historical origins of truth in metaphor and human affairs have worn away and been forgotten (Nietzsche 1968, 47). The origins of truth have been covered over by the realist rhetoric and political power of scientists and philosophers. With this realization, the idea of truth, as well as the distinction between truth and lie, becomes diluted.

The political variability and rhetorical character of truth led Nietzsche to conclude that all philosophical and scientific efforts at establishing transcendental, extrahuman, or objective knowledge were in vain. In response to projects such as positivism, which sought to uncover and ground knowledge in pure facts, Nietzsche (1968, 267) responded, "No, facts are precisely what there is not, only interpretations." Science and philosophy's attempts to apply an objective rationality to the world in order to establish indisputable facts were misguided and foolish. Their neutral vocabulary and earnest motives secretly hid an insatiable will to power. Nietzsche (1976, 42) wrote:

> In the great whirlpool of forces man stands with the conceit that this whirlpool is rational and has a rational aim: an error! The only rational thing we know is what little reason man has...One might invent such a fable and still not have illustrated sufficiently how wretched, how shadowy and flight, how aimless and arbitrary, the human intellect appears in nature...It is human, rather, and only its owner and producer gives it such importance, as if the world pivoted around it.

The comprehension of the world by a free-floating rational subject was for Nietzsche one of the great myths of modernity. Nietzsche (1968, 267) argued that rational subjectivity was not an a priori condition of humanity, but a constructed fiction that was "added and invented and projected behind what there is." The Cartesian subject, rationally capable of comprehending the world as it really exists, was a rhetorical construct of a particular cultural and linguistic perspective. It is part of "our grammatical custom that adds a doer to every deed" (Nietzsche 1968, 268).

Nietzsche was critical not only of rationally based theories of knowledge but also of all attempts in natural philosophy to ground knowledge in the empirical. For early experimentalists, the senses could be disciplined through method and the use of instruments in such a way that accurate representations of reality were possible. Nietzsche (1989, 247), however, argued that "the habits of our senses have woven us into lies and deception of sensation." From these lies there was "absolutely no escape, no backway or bypath into the real world" (Nietzsche 1989, 247). Sensory experience always and necessarily limited knowledge, for it was always derived from the perspective of the "perceiving being" (see Babich 1994, 89). The use of instruments to hone the senses could not counter this perceptual bias since they too relied on sensation.

Since foundational truth and transcendent rationality were fictions, those individuals and disciplines who advocated them should be seen as myth makers rather than truth makers. "They are all of them lawyers...and sly defenders of their prejudices which they christen 'truths'" (Nietzsche 1955, 5). Truthfulness was merely a moral value that was first applied to humans and only later to their actions (see Nietzsche 1955). This being the case, philosophy and science were to be seen as "human, all too human" activities, where anthropomorphism, perspectivism, interpretations, politics, and rhetoric, rather than the neutral uncovering of pure facts, prevailed. Even physics, the exemplar of scientific method and foundational truth, was said by Nietzsche to be "only an interpretation of the universe and an arrangement of it (to suit us, if I may be so bold!), rather than a clarification" (Nietzsche 1955, 15). The concepts of science and philosophy were merely products of faith that allowed people to endure life. Such a view made all knowledge a product of perspective and power relations. What made some knowledge claims appear truer than others was not their refined methodology or superior access to reality, but their political power. The authority to speak the truth came from the manifestation of a will to power and not the authority of nature speaking through the neutral voice of philosophy or the experimental sciences. Beneath the liberal and progressive masks of the modern episteme, Nietzsche found the ever-present elements of rhetoric, power, and an overly grand conceptualization of the potency and range of the human intellect.

Nietzsche's critique of rationality and foundational truth were not just directed toward what he considered to be modernity's epistemological fallacies. He was also deeply concerned with the effects of scientific rationalism, objectivity, and other "decadence values" on the conduct of everyday life (see Nietzsche 1976, 572).[1] Nietzsche argued that the endless modern drive for objectivity, which had its roots in the asceticism of Christianity, would finally culminate in the arrival of the "last man" (Nietzsche 1976, 129). The "last man" was a creature so permeated by an analytical sterility and the will to be objective and neutral that "he is no longer able to despise himself" (Nietzsche 1976, 129). In other words, the modern drive to be objective had produced or would produce an unanimated individual who could no longer feel or act with any type of conviction. All life had become rationalized to the point that every aspect of existence was open to the neutral, cold, and emotionless gaze of pure rational analysis. Consequently for Nietzsche, modern epistemology was not only misguided, but existentially dangerous: it threatened the spirit of humankind.

From Nietzsche's perspective, the deaths of God, foundational truth, and absolute morality provided no reasons to mourn. Instead their demise allowed individuals a sense of existential freedom or play and gaiety. There was no longer a need to adhere to old moral or epistemological foundations. In his words:

> Indeed, we philosophers and 'free spirits' feel, when we hear the news that 'the old god is dead,' as if a new dawn shone on us; our heart overflows with gratitude, amazement, premonitions, expectations. At long last the horizon appears free to us again once more...our ships venture out again, venture out to face any danger perhaps there has never yet been such an 'open sea' (Nietzsche 1974, 280).

Once truth was recognized to be a power-based illusion, individuals were freed to pursue their own methods and truths. Individuals were free to become "accomplished nihilists" who were capable of welcoming and celebrating the devaluation of the highest values (Vattimo 1988). The death of foundationalism permitted experimentation with new modes of knowing and living that were free from the confining restrictions of modern religion, epistemology, and morality.

Nietzsche's philosophical legacy clearly leaves twentieth-century theories of knowledge with some important programmatic problems. In following his advice, theory basically has two avenues: it can either accept a weak form of his revelations regarding the power and play of signification and become a form of rhetorical studies, or it can fully accept his critiques and become poetry. Neither option allows theory to offer precise, foundational statements on reality or to provide absolute truth. The first option makes disciplinary knowledge possible but curses it with contradiction and paradox. The second option escapes the paradox by making every form of writing into poetics and rhetoric but provides disciplinary problems when one wishes to elevate his or her claims over others. In some respects, much of twentieth-century social and cultural theory can be seen as attempts to engage and expand on Nietzsche's arguments about the "anthropological" limitations on truth.

Derrida, the Metaphysics of Presence, and Deconstructionism

In the late twentieth century, the Nietzschean critique of foundational truth has become influential in the establishment of poststructuralism, semiotics, and deconstructive literary and cultural criticism, or what has come to be generally labeled as "postmodern theory." Within this time frame, one of the most influential figures in the movement has been Jacques Derrida. Derrida's critiques of traditional philosophy and his deconstructionist strategy of reading texts have come to challenge a number of modern epistemic principles and precepts. Specifically, Derrida has reexamined four of the modern episteme's interrelated propositions: the autonomy of subjectivity and authorship, the promises of a metaphysics of presence, the possibilities of logicoempirical methodology, and the prospect of establishing correspondence referentiality.

Derrida begins one of his most influential works, *Of Gramatology* (1976), by stating that the problem of language and representation "has invaded the global horizon of the most diverse researches and most heterogenous discourses" (1976, 6). The problem of representation, as described and analyzed by Nietzsche and twentieth-century linguists and philosophers such as Saussure and Heidegger, had by the early 1960s spawned a "linguistic turn" (Rorty 1967). This linguistic turn made a dramatic impact on almost every domain of inquiry in the humanities and social sciences. Particularly affected by the renewed focus on language and representation were the areas of philosophy and literary theory. It would be in these disciplines that the postmodern critique of modern approaches to knowledge would become the most established and influential.

Previous views of language found in structural linguistics and correspondence versions of epistemology sought to make explicit the systems of relationships embedded in language. In structuralism, the goal was to identify the basic structural elements that made speech or culture possible. It was thought that such a move allowed for more precise analyses of the origins of culture. In correspondence theories of truth, the goal was to make language more opaque and representational. In the work of Derrida, however, both projects are viewed as severely misguided. For Derrida, both structuralism and correspondence referentiality are guilty of presenting the signifier and signified as being clearly distinguishable binary elements of the linguistic sign. Derrida felt that signifiers and signifieds are so closely intertwined that it is impossible to draw a distinction between these two aspects of the sign. This means that language is much more unstable and problematic than had been recognized by structuralists, correspondence epistemologists, or other cultural theorists. The basic elements of a language (or a culture) cannot be clearly defined since they are in continuous movement and transformation. In addition, and perhaps more important in Derrida's overall philosophical project, he argues that meaning is never fully present in a sign. The sign, which is marked solely by its difference from other signs, is influenced by the traces of the other signs that are absent from the original sign. The sign "is always inhabited by other signs which do not present themselves as such; there is always a deferral to something absent" (Spivak quoted in Pheby 1988, 57). Therefore, meaning is dispersed along a chain of signifiers (see Sarup 1989). "This interweaving results in each element being constituted on the basis of the trace within it of the other elements of the chain or system" (Derrida 1981, 26). This being the case, there is no such thing as stable meaning; "only, everywhere, differences and traces of traces" (Derrida 1981, 26).

The implications of this view of the sign and representation are vitally important for reexamining the modern construction and understanding of truth, particularly as manifested in scientific realism and accompanying correspondence theories of truth. Despite the insistence that the true meaning of an object or a text can be summoned before the rational, investigating subject, Derrida argues that meaning is always deferred or suspended. Language is incapable of reflecting reality since it is historical and culturally constituted and ever changing. In stronger terms, it is possible to argue that language creates and establishes the thing called reality. Since the sign is differential and filled with the traces of other signifiers, meaning can never be fully present to consciousness. Meaning is always culturally and linguistically mediated. The modern search for a metaphysics of presence that would summon the object of investigation directly to consciousness is and will always be a mistaken undertaking.

Since most of modern Western thought has seen language as nonproblematic, opaque, or at least correctable, Derrida's view allows for a reexamination of the entire project of modern epistemology. It now becomes possible to deconstruct the truth claims of authors by analyzing how they employ rhetorical and self-privileging strategies in the construction of their truth claims. In the words of Derrida (1981, 41),

> In a classical philosophical opposition we are not dealing with the peaceful coexistence of a vis-a-vis, but rather with a violent hierarchy. One of the two terms governs the other (axiologically, logically, etc.), or has the upper hand. To deconstruct the opposition, first of all, is to overturn the hierarchy at a given moment.

The goal of this type of deconstructionist reading is to "transform concepts, to displace them, to turn them against their presuppositions, to reinscribe them in other chains... and thereby to produce new configurations" (Derrida 1981, 24). In the process of deconstruction, the privileging of one term in a philosophical hierarchy is revealed as being more a rhetorical strategy of presentation than an accurate representational unveiling of some reality out in the world.

Derrida uses such a strategy of reading to deconstruct one of the basic hierarchies in modern Western theory, the division between speech and writing. In this division, Derrida critiques one of the oldest quests of philosophy, making objects directly present and accessible to consciousness. Within the Western philosophical tradition, speech has been viewed as more immediate and, hence, representational of the reality of presence. On the other hand, writing has been viewed as secondary and a merely second-hand derivative of reality. Speech directly accesses reality, while writing is merely a modification or distortion of reality. Within this standard modern view, speech provides the most reliable means to consciousness and truth since it represents immediate presence and a direct pathway from thought. In the Western tradition, writing is seen as "matter external to the spirit, to breath, to speech, and to the logos" (Derrida 1976, 35). In this logocentric view, "the voice is consciousness" (Derrida 1981, 22). In Derrida's approach, however, the positioning and privileging of speech over writing is indicative of Western thought's incessant search for immediate presence and absolute truth. This Western logocentrism represents a clear case of the way in which the privileging of one term over another serves as the foundation of a complete system of knowledge (e.g., rationality over irrationality). It is guided by the Western tradition's metaphysics of presence. Yet, such a task is always confounded by the slippery nature of language and representation, although this slipperiness has always been covered over by the rhetoric of philosophy. Writers are constantly struggling to repair the damage language does to their truth claims.

While most of Derrida's deconstruction has been performed on philosophical texts, it arguably has had its greatest impact on disciplines that examine literary texts. For example, in literary criticism one of the effects of deconstruction has been a complete reexamination of the interpretative meaning of literary works and arguments. Deconstructionism shows how the established hierarchy of a text eventually undermines the work's initial rhetorical strategy. In this process, the rhetorical play and conceptual privileging of the work is brought to the forefront for analysis. In literary criticism, deconstructionism has had the effect of removing the author as the primary vehicle for uncovering a text's true meaning. Thus, in a broad sense, the interpretation rendered by a literary critic is as valid as the author's so-called original intentions. "Original meaning" is both unimportant and impossible since it is no closer to the true story of the text than the comments and interpretations of the literary critic. Literary critics, like all writers, can only uncover traces and more traces, never original meaning. The search for origins only leads to more and more language. Consequently, textual criticism should be recognized as always an interpretative event and a contextually contingent accomplishment. One can never know the true meaning of a text since meaning is constantly shifting and being reinterpreted. Likewise, situating a text in its historical or cultural setting in order to extract its true meaning, as in traditional forms of literary criticism, is also an impossible task. History and culture are not neutral objects that exist "out there," but are themselves human discursive artifacts.

One of the most important implications of deconstructionism for epistemological inquiry is its leveling of any privileged access to truth. It therefore can be seen as initiating a type of "representational democracy." If the play of difference in the sign makes meaning impossible to "nail down" or summon before consciousness, then all forms of communication, including science and philosophy, can be seen as textual and discursive creations. Consequently, philosophical or scientific texts are as fictional as literary works. In fact, the long-held binary distinction between fiction and nonfiction becomes erased. This erasure prohibits the elevation of philosophy, social science, or science over literature or literary criticism as superior or higher forms of discourse. For Derrida (1976, 158), there is literally "nothing outside of the text." External reference is always a matter of intertextuality (see Baynes, Bohman, and McCarthy 1987, 122). No form of discourse, not even the serious speech of professional knowledge producers, is capable of escaping the interweaving of language to find and report an unmediated reality. Human discursive models and configurations allow the individual to see only certain things (see Jameson 1972, 14). Discursive models never reflect reality itself. All we have are epistemologically symmetrical discursive formations or texts and not the fully present thing or idea. As a consequence, objective knowledge about the world, as dreamed in the modern episteme, is impossible. Representation is revealed to be an ideology and foundationalism a myth (see Boyne and Rattansi 1990, 12).

What is the outcome of all this deconstructing? Is Derrida offering an alternative model to traditional epistemology? Or, have we, as one interviewer of Derrida once remarked, been led into a labyrinth (Ronse in Derrida 1981, 5). Derrida and his followers have repeatedly claimed that he is not offering a new system of knowledge, that, for example, promotes absence over presence or writing over speech (see Agger 1994). In fact, Derrida argues that he has a great admiration for the works he deconstructs (see Derrida 1981, 15). Rather, proponents argue that his work attempts to illustrate "the necessity of regarding the distinction of signifier and signified as functional and provisional rather than substantial" (Culler 1986, 144). In other words, his works help in revealing that the conceptual hierarchies that make a system of knowledge possible must be recognized as contingent and constructed and not grounded in an ultimate state of the world. Once we begin to unravel this hierarchy, we can better appreciate and critique the cultural work and power elements used to make and sustain "reality."

Deconstructionism, then, is not a new method for interpretation, at least according to its proponents: it is not "a specialized set of discursive procedures" or "a new hermeneutical method" (Derrida in Madison 1993, 216). In fact, it is extremely skeptical and reflexive about such endeavors. As Stanley Fish (1989, 493) has remarked, deconstructionism "continually uncovers the truth of rhetorical operations, the truth that all operations, including the operations of deconstructionism itself, are rhetorical." It argues that "to recognize the importance of language is to do so within the language" (Lawson 1985, 9). This being the case, there is "no time out from interpretation" (Fuchs and Ward 1992, 482). Interpretation is an ongoing process filled with constant revisions and renegotiations. In this process there can never be just one meaning of events or objects since these interpretations will always be made from a historical or cultural perspective. What is more, once events are transformed into texts, the instability of language renders such texts forever unstable. In such a setting, all knowledge producers can do is pile interpretation on top of interpretation or text upon text.

Baudrillard and the End of the Social-as-Real

The postmodern attempt to textualize truth can be seen in one of its more extreme versions in the work of Jean Baudrillard. It is perhaps Baudrillard, above all others who are labeled postmodernist, who pushes the collapse of correspondence referentiality and foundationalism to its ultimate nihilistic conclusion. Baudrillard begins his analysis with a somewhat straightforward social theory of the end of word/world correspondence; however, his conclusions lead to a literary-based, postrealist epistemology (Gane 1991, 167–173). In this post-realist epistemology, the captured object of the modern episteme escapes from the control of the rational cogito and returns to the "fatal destiny of the object itself" (Gane 1991, 167). As we shall see, his arguments are at once analytical and polemical; realist and antirealist. It is this "code switching" and the demolition of traditional theoretical distinctions that make Baudrillard's brand of postmodernism congruent with, although perhaps more extreme than, the orientation of Lyotard, Derrida, Foucault, and others who have been labeled postmodernists.

Baudrillard's "social theory" of the end of correspondence referentiality begins with an exploration of the proliferation of the commodity sign in the simulational world of television and other informational technologies. The proliferation of the sign has completely permeated the sphere of language, communication, and, as a consequence, society. Postindustrial consumer society is so "entangled in illusion that the distinction between true and false, TV and reality, cannot legitimately be made" (Best 1989, 38). With this proliferation and blurring of original notions of direct signification caused by the various informational technologies, the real has become the hyperreal. "We have become completely absorbed by models, completely absorbed by fashion, completely absorbed by simulation" (Baudrillard 1988a: 187). As a result, traditional notions of referentiality or correspondence have collapsed. The "precession of the model...puts an end to the real" (Baudrillard 1983, 98–99). It is impossible to establish correspondence referentiality in a social climate where the sign replaces and obliterates the object it is said to represent. It is not simply that reality has been replaced by simulacrum; rather, it is no longer possible to distinguish between the two.

One important result of the collapse of referentiality is that that which we once called society has become an amorphous and indistinguishable "mass." Social distinctions such as culture, ethnicity, and class, like philosophical distinctions, have become leveled by the homogeneous and homogenizing hum of television and related simulational technologies. In Baudrillard's (1983, 20) words, "the masses are no longer a referent because they no longer belong to the order of representation." The masses as a signifier no longer correspond to an object-in-the-world. The masses no longer can be referred to as a civil society, culture, "the people," class, or a repository for revolution or social action. All referentiality, and with it all attached meaning, has been leveled. With this leveling comes the "implosion" of the social-as-signifier. The social has become empty. Baudrillard (1983, 66) writes:

> If the social is both destroyed by what produces it (the media, information) and reabsorbed by what it produces (the masses), it follows that its definition is empty, and that this term which serves as universal alibi for every discourse, no longer analyzes anything, no longer designates anything. Not only is it superfluous and useless...it conceals that it is only abstraction and residue, or even simply an effect of the social, a simulation and an illusion.

The social as a "universal alibi" for politicians, humanists, and social scientists has ceased to exist. The social as a grand concept or reality principle has been destroyed by information technologies and redeployed in an undifferentiated and indistinguishable mass. Here, the social ceases to have epistemic foundations, hermeneutical meaning, or critical potentiality.

> (The masses) don't express themselves, they are surveyed. They don't reflect upon themselves, they are tested. The referendum (and the media are a constant referendum of directed questions and answers) has been substituted for the political referent. Now polls, tests, the referendum, media are devices which no longer belong to a dimension of representation, but to one of simulation. (Baudrillard 1983, 20)

A sociality with any inherent meaning has ceased to exist and is replaced by the everpresent fatal play of surfaces.

For Baudrillard, the rise of the undifferentiated mass marks not only the end or implosion of the social, but also the end of that positivistic or realistic discourse that developed and sought to capture it—sociology. Sociology, as the master discourse on the social, no longer has an object, a point of reference, a signified. Its epistemological foundation in a social realism has been eroded. Therefore, sociology dies alongside the social. Baudrillard (1983, 4) writes:

> Sociology can only depict the expansion of the social and its vicissitudes. It survives only on the positive and definitive hypothesis of the social. The reabsorption, the implosion of the social escapes it. The hypothesis of the death of the social is also that of its own death.

The social and sociology have been swallowed up by a new version of society. In this version all transcendence and referentiality are impossible. In other words, it is no longer possible for the empirically oriented sociologist to utilize the term "society" to capture the social. Speaking as a sociologist (or, more accurately, as an antisociologist), Baudrillard (1988a, 210) writes:

> The situation no longer permits us to isolate reality or human nature as a fundamental variable. The result is therefore not to provide any additional information or to shed any light on reality, but on the contrary, because we will never in the future be able to separate reality from its statistical, simulative projection in the media, a state of suspense and definitive uncertainty about reality.

Sociology then dissolves into a form of literature, poetics, or science fiction—although the "gibberish of the 'social sciences' will still echo it (i.e., the social) long after its disappearance" (Baudrillard 1983, 68). It no longer can claim a privileged epistemological position from which to express and defend its truth claims. Sociology's epistemological position was based on the ability of its perspective and concepts to extract meaning from (or impart meaning to) the socially real. These meaning-extracting and-granting functions of sociology are no longer possible in an era marked by the social's disappearance or implosion.

Ironically, Baudrillard's argument is a "social theory" of the death of social theory and all correspondence, epistemologically based forms of argumentation. It is meant to be the final word on the irrelevance of sociology and other realist-based theories in the contemporary world. Herein lies Baudrillard's postrealist epistemology. Henceforth, Baudrillard commits himself to writing something akin to science fiction or an "absurdist pataphysics" (Best 1989, 36). "Since the world drives to a delirious state of things, we must drive to a delirious point of view" (Baudrillard quoted in Gane 1991, 7). Baudrillard's poetic science fiction may have "flashes of insights," but it is more akin to literature rather than to traditional critique, analysis, or theory. In his view, the only option of theory is to become excessive and exuberant. Theory must accept the fatal and erratic destiny of the object, including itself. Baudrillard writes (1988b, 98):

> (Theory) must become excessive and sacrificial to speak about excess and sacrifice. It must become simulation if it speaks about simulation, and deploy the same strategy as its object. If it speaks about seduction, theory must become the seducer, and deploy the same stratagems.

Theory's only option is to become like the simulation model. It must forget the search for a deep logic or underlying code and relegate itself to the play of surfaces and exorcism. Theory can only "defy the world to be more: more objective, more ironic, more seductive, more real or more unreal…it has meaning only in terms of this exorcism" (Baudrillard 1988b, 100). Theory must try, in a Nietzschean manner, to recognize the masses "as the repository of a finally delusive, illusive, and allusive strategy, the correlative of an ironic, joyful, and seductive unconscious" (Baudrillard 1988a, 217).

For Baudrillard, not only is the individual decentered, as structuralism, poststructuralism, and sociology have long argued, but so too is its replacement—the social. It seems that the social is as unstable as "modernist man." In Baudrillard's view, the social is now a text without depth. It is nothing, and never has been anything, but a system of signification. When this signification becomes confused or blurred, the social simply ceases to exist—it becomes empty and implodes. There is no longer a correspondence between the theoretical word and the social world.

The radical nature of Baudrillard's brand of postmodern theory can be found in remarks he made during the early 1990s with regard to the Persian Gulf conflict. In an article in *The Guardian*, Baudrillard (1991) declared confidently that the war would never happen. From his view, the entire episode was a product of media simulation, public relations, and the management and manipulation of public opinion (see Norris 1992, 11). The Gulf War exhibited all the signs of postmodern hyperreality: the "reality" of the conflict could not be separated from the manufactured signs and media simulations that accompanied it. In fact, the omnipresence of manufactured signs made it impossible to talk about a real war at all. The conflict was a Hollywood-produced, made-for-TV movie complete with agents, scripts, actors, marketing departments, special effects, and directors.

The war was unreal: a type of war "without the symptoms of war, a form of war which means never needing to face up to war..." (Baudrillard 1991). For Baudrillard, the Gulf War was entirely rhetorical and simulational. It was further evidence that we live in a time and place where the division between truth and simulacrum is impossible to make. In turn, all those historical projects of modernity that sought to sort the representational from the real (e.g., Plato, Kant, Hegel, Marx, etc.) are completely and utterly dead (see Norris 1992, 15). The historical project of truth has become obsolete.

One perhaps can see still a bit of critical theory left in Baudrillard's views. His, however, is more of a "theater of the absurd" or affirmative Adornesque approach to critical theory rather than a traditional critique of society. Within Baudrillard's nominalistically influenced framework, if referentiality is no longer possible, then all hopes of finding fixed, determinate meaning, such as in the social, are in vain. The only courses of action that remain are nihilistic surrender, guerrilla warfare, or fatalistic poetics. Within the context of his views, traditional realist theorists are, at best, bad poets or, at worst, unsophisticated con artists.

Essentially, Baudrillard's writings tell us that the "story of the story of stories cannot be told" (Second of January Group, 1986, quoted in Ashmore 1989, 69). No epistemologically based narrative can obtain the true. All we have are simulations of reality—no reality, only hyperreality. Baudrillard's response to this postmodern condition is "the initiation of a style of (anti-) sociology as prose-poetry fiction theory" (Gane 1991, 131–132). It marks the final and most radical culmination of the project began by Nietzsche almost a century earlier.

Conclusion: The Politics of Rhetoric and Reality

Lyotard (1984, xxiv) has summed up postmodernism's antifoundationalist view of knowledge by defining it as an "incredulity toward metanarratives." In his view, the various discourses that support modernity, such as "the dialectics of spirit" and "the hermeneutics of meaning" have now lost their legitimacy and credibility (Lyotard 1984, xxiii). Our postmodern condition has revealed "a truth about our basic condition that has been covered over by comforting stories about the cosmos or history" (Kolb 1986, 258). For Lyotard, all discourse and truth claims are heteromorphous. They contain radically different language games with their own unique internal linguistic rules, moves, and structures (see Kellner

1988, 250). In this heteromorphous environment no language game or discourse is capable of grasping and reporting reality. Consequently, "all we can do is gaze in wonderment at the diversity of discursive species..."(Lyotard 1984, 26). We do not need nor can we support any type of archetypal theory or grand narrative that unites or overcomes these forever heterogeneous language games.

It is in this sense that postmodernism can be viewed as instituting the "turn against theory and toward narrative" now so evident in contemporary intellectual culture (Rorty 1989, xvi). For postmodernists, the concept of "theory" is a product of the modern episteme. It is part of the modern epistemic idea that an accurate account of phenomena can be found, conceptualized, and brought to light. The concept theory signifies the modern drive to forever nail down meaning. Narrative, on the other hand, suggests a storytelling or discursive enterprise. Narrative makes no pretenses about speaking the truth, for such an attempt is merely another order of storytelling. In this key move of postmodern theory, all knowledge and truth claims become textualized, storytelling activities. Everything becomes rhetoric and discourse. Neverending interpretation, not forever fixed theory or methodology, is all that is possible.

It is evident from the above discussion that much of postmodern theory adopts a strong Nietzschean stance with regard to the end of epistemic foundations. When truth is revealed to be an outcome of power configurations (Foucault 1977; 1980), linguistically biased and arbitrary philosophical hierarchies (Derrida 1981; 1982; Rorty 1979), or self-referential language games (Lyotard 1984), there is a tendency for postmodernists to conclude that modern conceptualizations of truth and reality are outmoded concepts. Since both truth and reality appear to be the products of discourse, there is little need to write as if discourse must originate in them. The only viable option open for theory, or what is now better written as "theory," is to recognize itself as a form of literature and practice poetics or polemics. In the literary model, "theory" no longer needs to defend its claim of representing the real—for the real is always brought into being by language (see Rorty 1982). The end of realist epistemology means theory is free to pursue a type of literary or textual criticism rather than science and analysis. The end of epistemology means theorists are free, like poets and literary critics, to experiment with new, nonrepresentational discourse—to pursue a type of poetics or musicology rather than science and so-called rigorous analysis (see Rorty 1982; Habermas 1987, 185–219).

From the postmodern perspective, any historical attempt to break out of literature and subdue language, like those found throughout the modern episteme, folds back on itself and "engenders the repetition and continuation of literature" (de Man 1971, 162). All reality is shaped and must pass through language and representation. From this view, "all significant forms of order may be reduced to a language like process" (Reiss 1988, 4). Such a view of representation often leads postmodernists to make claims such as that the Gulf War is a fictional hyper-reality or that "South Africa and Apartheid are, like you and me, part of [the] general text" (Derrida in Madison 1993, 216). All the world and its objects are representations. There is no cultural or historical context in which a text may be situated to reveal its truth either, for these too are just other texts. Simply put, we never have unmediated reality, only texts or chains of signs about reality. Representations are no longer just innocent signs, but "practices that systematically form the object of which they speak" (Foucault 1972, 49).[2]

In very simplified terms it is possible to conclude that the modern approach to knowledge is outward looking. It is primarily concerned with providing explanations of things. In contrast, the postmodern approach to knowledge is much more inward looking. It is concerned (modernists would say obsessed) with explaining the act of explaining. Instead of trying to develop strategies for creating accurate knowledge, postmodernists question the entire idea of accurate knowledge production. In postmodernism, knowledge turns on itself. The goal of investigation moves from providing knowledge about the world to providing an appraisal of the knowledge of it. In the end the postmodernists want to force knowledge producers to realize the forever tentative and political nature of all that is said and written.

When examining some of the central tenets of postmodernism and the general modernism/postmodernism debate, it is important to keep in mind the political undercurrents that have shaped these tenets and debates. While the topics unleashed by the postmodern textualization of truth have created considerable controversy between so-called modern realists and postmodern relativists and have spawned important questions about what intellectuals are now supposed to be doing, it would be a mistake to view the debate as just an ontological or epistemological conflict between competing views of reality and truth. The debate can also be seen as a deeply rooted political struggle over intellectual property and knowledge production and legitimization (Ward 1995b). Just as advocates of the New Science attacked the humanist "defenders of the text" in the seventeenth century, postmodernists attack the scientific "defenders of the real" in the twentieth century. Since its development in the sixteenth and seventeenth century and particularly since its institutionalization in the nineteenth century, science has held both an epistemically and politically dominant position in society and academia. The boundary conventions developed within the modern episteme have been used to relegate the humanities to a peripheral position in knowledge production. From the view of science, the humanities might produce interesting literature, poetry, and essays and provide a greater appreciation of life; however, they are not methodologically equipped to answer important questions regarding what is true or false or what is real or illusory. Thus, despite some problems, "science is the best procedure yet discovered for exposing fundamental truths about the world" (Atkins 1995, 97). Postmodern theory can be seen as an attempt to end or curtail that epistemic and academic hegemony. Its introduction of epistemic relativism and representational problems serves to challenge the scientific monopoly over truth. If the postmodernists are successful, scientists will no longer be the only group capable of making facts, although these facts are really better understood now as fictions.

Notes

1. A similar position can be found in Max Weber's discussion of the rationalization process in the *Protestant Ethic and the Spirit of Capitalism* (1952).

2. The political or practical outcome of postmodern theory is not entirely clear. Politically, postmodernists are divided between those who seek to use postmodernism to reconstitute leftist politics (see Agger 1993; Seidman 1991) and those that think that such a move only repeats the foundationalist mistakes of modernism (see Fraser 1993). In terms of academic practice, confusion reigns in many circles over what postmodern intellectuals are now supposed to be doing. On the one hand, it appears that if one follows the implications of postmodern theory, academics are required to completely abandon the truth game advocated in the modern episteme and its divisions between true and false. In this case, it appears that knowledge producers are to treat all knowledge, including their own, symmetrically and textually and become poets. Such an avenue can be found in projects that experiment with new modes of writing or alternative media. These projects, like postmodern art, use irony, multiple and conflicting voices, and recursive footnotes to play with, reinscribe, and problematize meaning. The second possibility allows academics to continue to play the game, but to play it modestly (see Fuchs and Ward 1994, 484). Within this scenario, theorists adopt a "minimal realist" strategy that abandons metaphysical certitude for more pragmatic ground rules of conduct. Here, so-called normal

methods of inquiry (e.g., surveys, historical research, experiments, etc.), may be used, but with caution and humility. Such a strategy can be found in the "rhetoric of inquiry" movement, which performs a type of discourse analysis on text using standard social science methods.

Chapter 3

Deconstruct(ed)ing Science

Within postmodern theory, all historical and ongoing attempts to establish a truth that corresponds with reality become suspect. More recently, these suspicions have extended beyond the projects found in philosophy and literature to include science and scientific realism, two of the most sacred institutions and philosophies of the modern episteme. As a result of postmodern theory's skeptical appraisal of all metanarratives and its emphasis on the multiplicity and indeterminacy of meaning, science too comes to be recognized as having a problem with representation. Postmodern theory suggests that the descriptive and explanatory language of science, like all languages, is fraught with problems of interpretation, translation, reception, and referentality. From this view, scientific representations can never be more than contingent and context-dependent approximations of reality. In stronger terms, representations are actually responsible for reality. Scientific representations are always and everywhere the product of language, metaphor, and localized semantic practices. This being the case, science can be viewed and critiqued as a storytelling enterprise—perhaps the most powerful storytelling enterprise of the twentieth century.

Until recently, postmodern-inspired critiques of knowledge were directed almost exclusively at literary, philosophical, or social scientific texts. In the last ten years or so, however, there has been a growing interest in using concepts from postmodern theory, particularly deconstructionism, semiotics, and Foucauldian archeology, to unveil the representational practices found in the natural sciences. In this chapter, I examine how the general postmodern critique of knowledge has been incorporated into analyzing science. It is not my contention that these postmodern analyses of science all use Derrida's deconstructionist theory verbatim or that they all reach the radical antirealist conclusions of Baudrillard. In fact, some of those who examine the rhetoric of science display a marked realist streak—they are what one might call weak textual relativist. Rather, I suggest that the general skepticism toward knowledge claims produced by postmodern theory makes possible the revival of interest in and use of rhetoric and literary methods to examine science. In short, postmodern theory allows rhetoric to come out of hiding and into the open. It makes rhetoric no longer a dirty word used to denounce and put certain types of claims in their place, but a type of reading that can be applied to all knowledge forms including rhetorical studies themselves.

Postmodern theory makes the restoration of rhetoric possible in three interrelated moves. First, it reintroduces the important role of language and representation in all knowledge-making practices. Second, postmodern theory levels or symmetricalizes various forms of discourse and truth claims. This move requires that all knowledge forms be treated as equally valid. Finally, it calls for a radical reflexive stance regarding its own knowledge (see Aune 1990, 254). This move prohibits even rhetoric itself from having any ultimate foundations. In these moves, postmodern theory makes rhetoric an integral part of coming to terms with the political orientation of all knowledge claims, as well as something one no longer has to be ashamed of doing.

In what follows, I focus first on how rhetorical and textual analysis have been used to examine the natural sciences. I explore how the so-called rhetorical turn has been employed to transform scientific practices and knowledge into texts open to literary criticism and rhetorical analysis. In this chapter, I also survey feminist critiques of science. These critiques often draw from literary methods of investigation and are given epistemological strength by the postmodern emphasis on perspectivism and multiple voicing. What unites the rhetoric of science and feminist critiques of science is the conviction that science is always affected by cultural or extracognitive forces.[1] These forces can be in the form of cultural representations or gender hierarchies. Finally, in this chapter I consider the recent offensive launched by scientists against postmodernism and feminist critiques of science. Specifically, I examine the counterattack found in the book *Higher Superstition* (Gross and Levitt 1994). To date, the work contains the most thorough scientific response to rhetorical criticisms of science.

Science as Rhetoric and Text

With all knowledge forms now potentially opened to literary analysis by postmodern theory, it has proven possible to develop a rhetoric of scientific inquiry: to look at the various metaphors, linguistic devices, syntactic options, narrative structures, interpretative actions, and semantic strategies scientists and other knowledge producers use to make and sustain their truth claims. This so-called "rhetoric of inquiry movement" argues that "the practice of inquiry might more usefully be understood in rhetorical terms" (Simons 1990, 2). It therefore can be seen as an attempt to counter "the dominant objectivist presuppositions of our age" (Simons 1990, 1). Within this movement, like with the deconstruction of literature and philosophy, scientific theories are often seen as stories or narratives. As with all stories, scientific theories are considered as "a translation or retelling from our lived-experience of everyday reality to a level of theoretical explanation" (Lewin 1994, 35). Therefore, scientific theories are to be inspected, like all texts, as a pragmatic attempt by a group of people to make sense of certain types of events and phenomena using established linguistic conventions. These conventions draw their descriptions from local text-making activities and established forms of rhetoric, and not reality.

One of the most influential recent attempts to analyze science using such literary methods can be found in Charles Bazerman's *Shaping Written Knowledge* (1988). In this book, Bazerman's central focus is the scientific research article. He considers the scientific article to be a unique literary genre with its own particular strategies of rhetorical persuasion. For Bazerman, the persuasiveness and power of the scientific article comes primarily from its concealment of the human action used to construct it. It silently employs what the sociologist Joseph Gusfield (1976, 17) once called a "style of non-style." Unlike novels or poetry, which often consciously draw attention to their own creation, the scientific article downplays the active, artistic role of the author. It consciously conceals the various contexts and interpretative marks of its creation in order to make the article appear to speak "univocally about nature" (Bazerman 1988, 27).

For Bazerman (1988, 26), in the process of writing science all scientific authors ask themselves how they can use the background of knowledge within the discipline to "present an original claim about a phenomenon to the appropriate audience convincingly so that thinking and behavior will be modified accordingly?" Thus, article writing always involves political motivation and calculated rhetorical invention. The author must consider the tactics to use in order to convince an audience of his or her truth claim. According to Bazerman, in producing a successful and convincing argument, the scientific author must weight and balance four mutually dependent contexts: the object under examination, the literature in the field, the audience, and the author herself (Bazerman 1988, 24). Within each context the scientist must develop usable rhetorical strategies that are concurrent and coherent with her particular linguistic community. For example, an overly reflexive scientific writer might be ridiculed or ignored because his or her work violates the group's own rules regarding the author's impact on an article. Likewise, an author may be maligned for not incorporating and encountering some of the classical works in a field. Texts are, then, to be viewed as solutions to the problems of context (Bazerman 1988, 26). They are conscious strategies that attempt to use symbolization to weigh and balance the various contexts under which writing occurs. Bazerman sees his goal as carefully showing scientists how "writing matters": how rhetoric and literary invention are key communal ingredients for doing and conveying science. He feels that if scientists and other knowledge producers can see how rhetoric and symbolization work in the writing process, they will be able to write better texts (see Bazerman 1988, 317).

Another important recent attempt to explore the rhetorical dimensions of science can be found in the work of Lawrence Prelli (1989). Prelli too focuses on the various rhetorical inventions used by scientists to put forth truth claims. He argues that science is a special kind of communicative collectivity. This collectivity is significantly different from those found in other communicative groups, such as religious and lay communities. This uniqueness means that scientists operate under their own distinctive rhetorical schemes that arise from the norms of the scientific community. He concludes that "doing science is not an exercise of formal logic or of following some neutral, procedural algorithm. Nor is it purely idiosyncratic or illogical" (Prelli 1989, 265). Instead, doing science is a rather localized, rhetorically influenced construction of a claim based on the interactive and interpretative rules of the scientific community. In order for a claim to be successful, it must always appeal to what is "expected in reasonable scientific argument with a technical rhetorical situation" (Prelli 1989, 260).

Both Bazerman and Prelli take what could be considered a moderate stance to the role of rhetoric in science. They are clear examples of the weak textual relativism discussed in the introduction. A more radical attempt to apply rhetorical analysis to science can be found in the work of Alan Gross (1990). Gross's undertaking is far more ambitious and far-reaching than Prelli's and much less normative than Bazerman's. It also quickly becomes evident that his approach is much more strongly influenced by postmodern theoretical currents. While Bazerman and Prelli (1989, 8) contend that there are some aspects of science that are beyond rhetorical analysis, Gross argues that science is thoroughly and inextricably rhetorical. In addition, while Bazerman and Prelli want, at least in some manner, to use rhetorical analysis to make scientific inquiry and writing more effective, Gross contends that such a goal is highly untenable. He sees the rhetorical analysis of science as part of a larger reassessment of all truth claims and established knowledge hierarchies. Gross argues that as rhetorical analysis proceeds, "science may be progressively revealed not as the privileged route to certain knowledge, but as another intellectual enterprise" (Gross 1990, 3). Such a leveling of the hierarchy between science and nonscience means that science must take "its place beside, but not above, philosophy, literary criticism, history, and rhetoric itself" (Gross 1990, 3). In the aftermath of rhetorical analysis, science will become just another way of talking about the world, not the only or correct way.

For Gross, science is a thoroughly rhetorical enterprise. Like all forms of discourse, science is by necessity "centered on persuasion" (Gross 1990, 6). As in everyday life, the use of scientific language always involves the desire of a person or group to sway others to the person's or group's point of view. The revelations of science appear not through "discovery," but exclusively through rhetorical and literary maneuvers and modes of presentation. Scientists, like all who use and manipulate language, utilize a host of rhetorical devices and strategies to persuade others of their truth claims. It is successful and persuasive story-tellers who, in turn, become successful truth makers.

Gross uses this view of language and rhetoric to deny vehemently the realist argument that scientific theories are direct representations of the "brute facts" of nature. He writes:

> In fact, in no scientific case do uninterpreted brute facts—stellar positions, test-tube residues—confirm or disconfirm theories. The brute facts of science are stellar positions or test-tube residues *under a certain description;* and it is these descriptions that constitute meaning in the sciences. (Gross 1990, 11)

Since all descriptions of the world must pass through language, rhetoric, and metaphor, no description can claim it is free from the effects of human interpretative or rhetorical intervention. It is this process of interpretative and rhetorical intervention that by necessity distorts the essence of the object being studied. This being the case, we never have "rhetoric on the one hand and science, or autobiography, on the other, but a fusion of the two" (Gross 1990, 65). In other words, scientific knowledge cannot be separated from culturally derived representations or from the historical position of the interpreter doing the representing. It is as culturally formed as any other mode of discourse.

In analyzing various forms of scientific discourse, Gross finds a number of rhetorical inventions at work. One of the most important for science is the illusion of unattached objectivity found in scientific prose. Unlike Bacon's and others' attempts to control rhetoric in order to have a value-free reporting of things as they are, Gross (1990, 15) argues that objectivity itself involves the "assertion of a value." The language of objectivity serves as a rhetorical stratagem to give the impression of denied emotion and dispassionate observation. Under the gaze of rhetorical analysis, however, objectivity is revealed to be merely "a nonrational appeal to the authority of reason" (Gross 1990, 15). Appeal to reason serves as a deliberate scheme to force all analytic decisions into the domain of science. Here, science can establish the ground rules and dictate procedure and the modes of assessment. Science's will to objectivity secretly "embodies a convenient myth, a myth in which, apparently, reason has subjugated the passions" (Gross 1990, 15).

Gross also argues that the tables, charts, and graphs of scientific discourse are also part of its rhetorical strategy. In his view, these visual representations serve two central purposes: First, they create a virtual witnessing of experience. Second, they strongly suggest direct, causal relationships between objects and events (Gross 1990, 74). In the first case, visual representations serve to add semantic weight to an argument. They imply careful research, detached observation, and apolitical motives. The second function serves to convince an audience of the efficiency and firmness of an explanation. It serves to rule out alternative explanations and delimit other influencing forces. Tables, charts, and graphs are appendices added to an argument in order to make it more persuasive and convincing. These visual representations draw upon the long-held cultural belief that numbers do not lie or misrepresent (see also Porter 1995).

Gross uses such a rhetorical analysis to examine one of the most important scientific arguments of the twentieth century, Watson and Crick's article on deoxyribose nucleic acid (DNA). Gross argues that DNA is a thoroughly rhetorical invention. Its reality "is an effect only of words, numbers, and pictures judiciously used with persuasive intent" (Gross 1990, 54). Gross argues that the original DNA article is so persuasive an argument because it "is harnessed in the interest of a particular view of life" (Gross 1990: 61). More specifically, the DNA argument succeeds because it appeals to underlying and deeply held psychological views of other scientists. It invokes and recreates a psychological code reconfigured from fairy tales.

Gross concludes that because science is a discursive and linguistic activity, it can never describe things-in-themselves. "Facts are by nature linguistic—no language, no facts" (Gross 1990, 203). Rhetorical devices, like metaphors, are a necessity of representing. As such, these rhetorical devices cannot be escaped or subdued. They must be brought to the forefront so they can be analyzed and understood. "The causal structure of the world" turns out to be not a reflection of reality but "a hypostatization of a set of social and linguistic practices" (Gross 1990, 82). Science may be good at predicting patterns of nature, but it is unable to offer unbiased explanations of those patterns. In other words, prediction has nothing to do with explanation or truth (see Gross 1991).

Despite important differences between Bazerman's, Prelli's, and Gross's accounts of the role of rhetoric in science, all agree that rhetoric plays a prominent role in the creation of scientific knowledge. Despite the efforts of advocates of the modern episteme to separate rhetoric from truth, these authors contend that such a complete separation is impossible. The modern epistemic argument that there is science on one side and rhetoric on the other "is, at the very best, inaccurate in significant cases" (Prelli 1989, 262). Of the three works examined here, Gross's pushes the revival of rhetoric to its more postmodern conclusion. His work can be seen as a reflection of the dauntless attitude that postmodern theory now gives to humanistic intellectuals. Rhetoricians need not be relegated to study just some minor aspects of scientific discourse or to co-conspire in cleansing science of distortions, but now can use a strong form of rhetorical analysis to look at the entire domain of science, including its long-held claim to represent and report reality accurately. Thus, rhetoric is not just a small part of scientific practice, but the primary way it creates knowledge.

Engendered Knowledge: Feminist Critiques of Science

Accompanying the treatment of scientific knowledge as rhetoric and text has been a multifaceted attempt to unravel the gender dimensions of scientific knowledge. This project can be seen as part of a larger effort to offer feminist critiques of traditional epistemology and, in some cases, to offer a uniquely feminist approach to knowledge (see Duran 1991). Feminist theories of science constitute a wide range of perspectives and approaches. They range from less controversial analyses of gender equality in science faculty and funding and gender representation in scientific fields to more contentious examinations of the historical engendering of knowledge and projects to build a feminist epistemology (see Rosser 1989). What all of these different approaches and perspectives share is the certainty that gender constitutes one of the, if not the, most fundamental divisions found in all societies and their institutions and knowledge systems. Since science is part of a society and its knowledge systems, feminist theory makes possible the exploration of "the myriad ways in which it, too, is structured by expressions of gender" (Harding 1986, 57). This being the case, it is held by many feminist critics of science that the concept of gender and related concepts such as gender hierarchies and gender politics can be used to illuminate some, if not most, of the basic workings and divisions of modern science.

As in the rhetoric of science, many of the writings found in feminist critiques of science and epistemology are directed at modern science's conceptualization and promotion of objectivity, as well as its closely related subject/object dichotomy (see Keller 1982; Mangena 1994). For scientific realists, objectivity is both possible and necessary for discovering the nature of phenomena and for reporting knowledge. Feminist critiques of science, however, point out that this type of objectivity is but a reflection of a patriarchal society and masculinist culture's will to dominate and control. Objectivity is not an inherent mode of knowing but "a characteristic which has been applied to beliefs, individuals, theories, observations, and methods of inquiry" (Longino 1990, 62). In other words, objectivity secretly masks science's gender politics. The objective gaze of scientists that makes a separation between the knower and known has served as a cover for their often male-centered way of organizing and representing the world. It has served as a means for males to force all of the world under their perspective and control.

In many feminist critiques of science the great terms of modernity, such as progress, emancipation, reason, and enlightenment, turn out to be mechanisms to silence minorities and to suppress other views and experiences. The alleged detached objectivity of science has served as a covert means to exploit and dominate others. As Elizabeth Fee (1983, 13) has put it, "we find that the attributes of science are the attributes of males; the objectivity said to be productive of the scientific is especially identified as a male way of relating to the world." Scientists have mistaken their own gender situatedness and male perspective for the universal. The male perspective has been transformed into the only perspective used to represent all of nature and humankind. In this process, scientists are implicated in political oppression and annihilation. In their pursuit of objectivity, they have always had to marginalize the voices and perspectives that got in their way, particularly those of women and people of color. Feminist critics concur with Gross's argument that objectivity is the assertion of a political value in the guise of neutrality and dispassionate observation.

Since scientific objectivity can be viewed as a male myth and as a political means for disavowing the experiential knowledge contributions of women and others peripheral social groups, feminist critiques of science contend that scientific knowledge has often been directly employed to promote "sexist, racist, homophobic and classist social projects" (Harding 1986, 21). This sometimes results from "faulty research design, overextension of data and general 'bad' science" (Rosser 1989, 7), but often it is simply the result of expanding a masculinist perspective to being *the* perspective: what is male has been generalized to all humankind. These masculinist assumptions have impoverished scholarship by "causing us to focus on certain problems to the exclusion of others, and they have led us to make particular interpretations when equally valid alternatives were available" (Gilbert 1991, 173). The gender biases contained within scientific research have served as a political means for suppressing other points of view and interpretations of nature. It has supported the domination of males over women and other groups in the political hierarchy of society.

Historical and contemporary studies of science find many examples of how gender politics intervene in the supposedly neutral representations of science. Schiebinger (1993, 40–74), for example, finds Linnaeus's labeling of humans under the classification of Mammalia to be a result of the movement to discourage the use of wet nurses in the eighteenth century. In addition, she finds received notions of heterosexuality and appropriate gender roles to be at the heart of early works in botany to classify and describe plant reproduction (Schiebinger 1993, 11–39). Donna Haraway's (1989) work shows how the discipline of primatology has been filled with sexist and racist stereotypes and representations. Other writers identify links between sociobiology and attempts to regulate the role of women in society (see Birke 1986); how studies of the brain and cranial anatomy have been used to separate and hierarchize men and women (see Bleier 1984); and how gynecology and obstetrics texts have been used as pronatalist arguments to persuade women to have children and to stay at home (see Findlay 1993). For feminist critics of science, these studies provide clear evidence that science is not free from the gender biases or politics of society. In fact, science always both reflects and creates these biases.

Such critiques of objectivity and the various examples of the historical engenderings of scientific knowledge have opened up the possibility of developing a feminist way of knowing—of generating a distinct feminist epistemology. In examining those attempts to develop a feminist epistemology, it is possible to identify three general stances. First, there are those who wish to use feminist theory and feminist critiques of science to cleanse science of its overt and subtle sexism. From this perspective, the epistemology and methodology of science are basically sound: they have only been misapplied. If science can become more cognizant of the effects of social hierarchies and gender biases, it can build better, less sexist forms of knowledge. Second, there are those who attempt to use an essentialistic view of women to provide an alternative to traditional epistemology—a gynocentric-based epistemology. From this position, women, by their very nature, possess unique and superior ways of knowing (see Daly 1978; Ruddick 1989). The inherent attributes of women, such as sensitivity and intuition, can be used as a starting point for constructing an epistemology that is void of masculinist biases and patriarchal control and, in actuality, closer to reality. Within this epistemology women must be recognized as the true harbingers of insight and knowledge.

The third position maintains that building a feminist epistemology requires more than a correction of the status quo or a turning of the tables on patriarchal knowledge. It actually requires a wholesale reassessment of the Western intellectual tradition. From this view, "women's distinctive social activities provide the possibility for more complete and less perverse human understanding—but only the possibility" (Harding 1989, 197). Women's marginalized status, as well as their social and embodied experiences, hold the possibility of building a knowledge system that does not rely on exploitation and domination performed in the name of scientific objectivity. The masculinist approach attempts to dominate and control, while the feminist, because of women's social experiences and positions, seeks to understand and emancipate. This perspective argues that the essentialistic thinking of proponents of a gynocentric epistemology only reverses and repeats masculinist assumptions and biases. Likewise, the argument that science can be cleansed of its sexism fails to recognize how deeply sexism and gender distinctions are interwoven into the very structure of scientific epistemology and methodology. From this position, what is needed is a complete rethinking of the project of epistemology (see Harding 1986; Code 1991; Duran 1991).[2]

The various attempts to build a uniquely feminist epistemology have recently led to a much more reflexivistic turn in feminist analyses of science (see Hekman 1990; Nicholson 1990). This turn results, at least in part, from a realization that feminist theories of women have unwittingly adopted masculinist standards of universality in their description of and theorizing about women. In this case, even the notion of a feminist epistemology may itself harbinger the male will to totalize, dominate, and control. A postmodern feminism calls for the recognition of differences between women and for a multiple voicing of women's different experiences. This move prohibits absolutist and universal claims about women to be put forward at the expense of diversity. This postmodern turn may also be seen as a result of a more reflexive focus found throughout much of contemporary cultural theory. This reflexive inquiry begins to call into question the hidden realist representational assumptions contained within feminist theory and critiques of science themselves. The addition of postmodernism makes the idea of critique, in the modern sense, exceedingly problematic. The idea of critique assumes a privileged position from which to launch such an analysis. Postmodernism argues, however, that inherently privileged positions are no longer sustainable. Within the postmodern approach, there appears to be little possibility of developing a truly feminist epistemology. This is so because in dealing with science feminism has had to rethink theorizing itself (see Harding 1986, 251).

Scientists Strike Back: The Response to Postmodernism and Feminism

Until recently the various deconstructions of science and scientific knowledge performed by rhetoricians, feminist critics, and other critics went basically unchallenged by scientists. Natural scientists, perhaps with their privileged epistemic and political status as a buffer or because they were busy doing other things, have gone about their laboratory lives without much consideration of what nonscientists have had to say about them or the practice of science. There is some evidence to suggest, however, that this inattention is now beginning to change. In a recent book, *Higher Superstition*, biologist Paul Gross and mathematician Norman Levitt take specific aim at a number of science critiques, including the "grotesque tendencies" found in postmodern critiques of science and "the academic industry" of feminist critiques of science (Gross and Levitt 1994, 108; 244).[3]

In broad terms, Gross and Levitt (1994, 72; 88) see postmodernism as a left-wing-inspired attempt to transform everything, including science, into stories or narratives. From their view, this move allows postmodernists to move knowledge to the home turf of those who are "clever with words, disposed to read texts with minute attention and to attend to the higher-order resonances of language" (Gross and Levitt 1994, 74). They see postmodernism as the amplification of a "narrow area of literary criticism or rhetorical analysis into a methodology for making judgements of the entire cultural spectrum" (Gross and Levitt 1994, 75). It attempts to transform all knowledge strategies into political and cultural wars. Gross and Levitt (1994, 106) hope that:

> the painful bolus of postmodernism will pass through the costive bowels of academic life sooner rather than later. Pass, of course, it will eventually. Keeping the hard sciences from contamination should not be impossible, provided that the scientists' resistance to jargonistic snow jobs is as high as it ought to be. In the meantime, unfortunately, the postmodernists will be out there trying to dominate every intellectual conversation.

For Gross and Levitt, postmodernism is an arrogant, narcissistic, and self-serving movement that engages science only at an abstract, theoretical level. Postmodernists never encounter the actual practices of scientists; they only condemn science whole-cloth. In Gross's and Levitt's view, only scientists are capable of examining and critiquing scientific theory and practice. All others are envious amateurs who have no idea as to what science is actually about or what it is trying to accomplish.

Gross's and Levitt's attack is directed not just at postmodernists, however. They are also highly critical of feminist critiques of science. They argue that feminist critiques portray science as the enemy of liberation and emancipation when in fact it is one of the great champions of freedom and justice. From the feminist perspective, science must be overhauled or overthrown in order to allow for a truly nonsexist social order. For Gross and Levitt, however, such a view lacks any objective proof of science's misrepresentation of the world. Women's studies programs and their political agendas that support antiscience ideas have a "sacrosanct status" in academia. This status makes them immune "to the scrutiny and skepticism that are standard for other fields of inquiry" (Gross and Levitt 1994, 110). They write, "how sad it is that it should now emanate from the scholarly halls of universities and reverberate there among intellectuals who inherit the Enlightenment" (Gross and Levitt 1994, 148). Because women's studies programs are protected by the shield of past oppression, they are viewed as immune from the usual standards of proof necessary to establish a claim.

Essentially, the authors argue that it is postmodernism, feminist critiques, and other "anti-science" movements that are politically constructed and motivated, not science (see Gross and Levitt 1994, 69). Gross and Levitt see postmodernism and feminist critiques of science, as well as other movements that analyze science, as part of a general leftist conspiracy in the humanities and social sciences to destroy all faith in the central institutions of modern society—of which science is one of the most central. Disappointed by the failures of the 1960s, the left now seeks vengeance on all the tenants of the Enlightenment. These antiscience advocates use "an unrelenting moralism" to cloak themselves in "political and social virtue" (Gross and Levitt 1994, 246). They are, therefore, part of a misguided crusade toward a rigid political correctness that attempts to promote and extend the political agenda of the left.

In their condemnation, Gross and Levitt portray scientists as the good and virtuous defenders of rationality protecting themselves from the onslaught of the evil and misguided people of postmodern and feminist irrationality. They accuse academic groups that critique science as being guilty of "intellectual dereliction" (Gross and Levitt 1994, 239). From now on Gross and Levitt advise scientists to be on guard against the erosion of scientific rationality wherever it may occur. Scientists are encouraged to attend seminars given by nonscientists about science in order to set the record straight. They are invited to scrutinize the tenure decision of science critics and evaluate the science education curriculum at their respective universities to make sure it has not been infiltrated by anti-scientists (see Gross and Levitt 1994, 252–256). Gross and Levitt's critiques goes so far as to argue that if the humanities faculty were to walk out of an institution such as MIT, that the science faculty could manage to put together a respectable humanities program. On the other hand, if scientists were to walk out, the humanists would be unable to carry on the science curriculum (see Gross and Levitt 1994).

What is particularly interesting about Gross's and Levitt's work is its rancor. Scientists who bother to encounter postmodernism or feminist critiques of science often find it hard to believe that anyone would question the epistemic superiority of science. For scientists, science is an unequivocal good. It might have occasional and isolated problems, such as plagiarism, fraud or the mismanagement of funds, but on the whole its contributions to the progress of humanity are immeasurable. From this view, science is a testimony to the power of the human mind to comprehend and transform the world. Such a view is backed by the philosophy of scientific realism. From this view, "science succeeds precisely because it has accepted a bargain in which even the boldest of imagination stands hostage to reality" (Gross and Levitt 1994, 234). Reality, then, serves as an "overseer at one's shoulder" (Gross and Levitt 1994, 234). Part of scientists' disbelief is that for them the doctrine of scientific realism is not just an "ideology," "text," or "discursive formation." Rather, it is a view that is concurrent with the way reality operates. For scientists, scientific realism is both an ontological and moral foundation. Natural scientists cannot be relativist about their findings, for those findings are always and everywhere mitigated and, in successful cases, caused by reality.

Conclusion: Reversing the Causal Arrows of Science

Until recently, science remained the last largely unscrutinized institution and set of practices and ideas of modernity. While other realms of knowledge and social institutions have been slowly opened to a relativistic investigation, textual deconstruction, and critical evaluation, science has remained largely untouched. Part of this avoidance has to do with the crucial role science has historically played in

2.3 Model estimation

As with any other functional regression model, estimation of the parameters of a functional multinomial response model is an ill-posed problem due to the infinite dimension of the predictor space. See Ramsay and Silverman (2005) for a discussion on the functional linear model. In addition, the functional predictor is not observed continuously in time so that sample curves $x_i(t)$ are observed in a set of discrete time points $\{t_{ik} : k = 1, \ldots, m_i\}$ that could be different for each sample individual. The most used solution to this problems is to reduce dimension by performing a basis expansion of the functional predictor.

A first estimation of the parameter functions of a functional multicategory logit model can be obtained by considering that both the predictor curves as parameter functions belong a finite space generated by a basis of functions $x_i(t) = a_i'\Phi(t)$, $\beta_s(t) = \beta_s'\Phi(t)$, with $\Phi(t) = (\phi_1(t), \ldots, \phi_p(t))'$ a vector of basic functions that generate the space where $x(t)$ belong to, and $a_i = (a_{i1}, \ldots, a_{ip})'$ and $\beta_s = (\beta_{s1}, \ldots, \beta_{sp})'$ the vectors of basis coefficients of sample curves and parameter functions, respectively. The sample curves basis coefficients will be computed in a first step by using different approximation methods as interpolation (data observed without error) or least squares smoothing (noisy data).

Then, the functional model turns to a multiple one given by

$$l_{is} = \alpha_s + \int_T x_i(t)\,\beta_s(t)\,dt = \alpha_s + a_i'\Psi\beta_s \quad s = 1, \ldots, S-1, \quad i = 1, \ldots, n,$$

with $\Psi = (\psi_{uv})$ being the $p \times p$ matrix of inner products $\psi_{uv} = \int_T \phi_u(t)\,\phi_v(t)\,dt$.

In matrix form each vector of logit transformations $L_s = (l_{1s}, \ldots, l_{ns})'$ can be expressed as $L_s = \alpha_s\mathbf{1} + A\Psi\beta_s$, $s = 1, \ldots, S-1$.

The estimation of this model will be carried out by maximizing the associated multinomial log likelihood, under each of the four different multicategory logits considered in previous section. In the case of baseline-category logits the log likelihood is concave, and the Newton-Raphson method yields the ML parameter estimates. For cumulative logits a Fisher scoring algorithm is used for iterative calculation of ML estimates. The adjacent-categories logit model is fitted by using the same methods for its equivalent baseline-category logit model. In the case of continuation-ratio logit models the simultaneous ML estimation of its parameters can be reduced to separate fitting of model for each different continuation-ratio logit by using ML estimation for binary logit models.

2.4 Principal components approach

The ML estimation of the functional multinomial regression model obtained by the approach in the previous section is affected by high multicollinearity what makes the variances of estimated parameter function increase in an artificial way. This has been proven for the logit model of binary response (Aguilera *et al.*, 2005) and for its functional version (Escabias *et al.*, 2004) through different simulated and real data sets. In this paper this problem will be solved by using as covariates of the multiple multinomial regression model a set of functional principal components of the functional predictor. An alternative way of avoiding excessive local fluctuation in the estimated parameter function would be to use a roughness penalty approach based on maximizing a penalized likelihood function (see Marx and Eilers (1999) for the functional regression model with binary response).

Two different FPCA of the sample curves will be considered after approximating such curves in a finite dimension space generated by a basis of functions. First, we will compute FPCA of the sample paths with respect to the usual inner product in $L^2(T)$ that is equivalent to PCA of the data matrix $A\Psi^{1/2}$ with respect to the usual inner product in $\mathbb{R}^p$. And second, we will perform PCA of the design matrix $A\Psi$ with respect to the usual inner product in $\mathbb{R}^p$ that is equivalent to FPCA of certain transformation of sample curves $x_i(t)$. The results set out in Ocaña *et al.* (2007) allow to demonstrate these equivalences between functional and multivariate PCA. Let us observe that both FPCA match when the basis is orthonormal.

Let Γ be a matrix of functional principal components associated to $x(t)$ so that $\Gamma = A\Psi V$ with $VV' = I$. Then, the multinomial logit model can be equivalently expressed in terms of all principal components as $L_s = \alpha_s \mathbf{1} + A\Psi\beta_s = \alpha_s \mathbf{1} + \Gamma\gamma_s$, and we can give an ML estimation of the parameters of the functional model (coordinates of $\beta_s(t)$) through the estimation of this one, $\widehat{\beta}_s = V\widehat{\gamma}_s$.

Then, we propose to approximate these parameters functions by using a reduced set of principal components. There are different criteria in literature to select principal components in regression methods. Escabias *et al.* (2004) compared in the functional binary logit model the classical one that consist of including principal components in the model in the order given by explained variability with the one of including them in the order given by a stepwise method based on conditional likelihood ratio test. In this work we will compare these two methods for different functional nominal and ordinal logit models. The optimum number of principal components (model order) will be determines by using different criteria based on minimization the leave-one-out prediction error or the leave-one-out misclassification rate via cross-validation.

The model will be tested by different simulated examples and applications with real data. It will be shown that the best parameter function estimation is given by the model that minimizes the mean of the integrated mean squared all evaluations of knowledge. Since its development, science has offered the possibility of forever deciding the merits of conflicting truth claims and producing absolute knowledge. In modernity, it has been science and scientific expertise that have served as a foundation and final court of appeals for most knowledge claims. This lure of scientific foundationalism has enticed intellectual movements as diverse as Marxism, quantitative history, and literary criticism. Postmodernism's general epistemological skepticism coupled with its radical reflexivism have now made these enticements and possibilities much more problematic than before. By reestablishing the role of rhetoric in the representations of science and by pointing out the gender politics of science, rhetoricians and feminist critics of science have launched an attack on science's claims to be the true gateway to reality.

Postmodern and feminist critiques of science attack the philosophical position on which science and scientific hegemony rest. Specifically, they attack science's cherished distinctions between truth and rhetoric, its attempt to construct a universal descriptive metalanguage, its arguments for the possibility of objectivity, and the unmediated revelationary power of experimentalism. In doing so, the rhetoricians and feminist critics have sought to establish noncognitive factors, such as rhetoric and gender, as key constitutive ingredients of scientific knowledge. Rhetorical analysis of science and feminist critiques of science have installed rhetoric and language or gender as obligatory points of passage in accessing all knowledge creations.

The result of these various deconstructions of science is that scientific truth becomes just another text/ideology utilizing particular signifying practices, rhetorical forms, and political motives. All knowledge and wisdom come to be recognized as rhetorical maneuvers (see Jacobitti 1989, 93). Scientific statements can no longer be thought of as "the neutral instrument of a triumphant content" (Barthes 1976, 10) but as contextual and contaminated cultural products. Within such an intellectual climate it is easy to see how social psychology can be deconstructed (Parker and Shotter 1990); how anthropology becomes part of an ongoing interpretive dialogue (Tyler 1987); how physics can be said to lie (Cartwright 1983); how biology tells stories (Myers 1990); and why the revelations of economics and sociology are viewed as fictions (Brown 1988; McCloskey 1985).

The response of scientific realism to postmodernism has only recently appeared on the scene. A response like Gross's and Levitt's, however, seems ill-equipped to handle postmodernism's epistemological and reflexivist challenges. All that scientific realists such as Gross and Levitt can do is try to rally the troops against an invasion into their sacred domain. Their indignation and incessant morality undoubtedly rings a bit hollow for postmodernists, however. Polemical works such as *Higher Superstition* (written by scientists who are not supposed to be polemical) ironically establish the postmodernist point that while scientific knowledge claims to provide objective and unattached observation, it is in the end thoroughly rhetorical and political. It only fuels the postmodern contention that we are indeed always in the middle of a knowledge and cultural war—that knowledge making is merely war by other means.

In making the claim that all knowledge is contextual, postmodern critiques of science can be seen as an attempt to reverse scientific realism's causal arrows: instead of reality being reflected in accurate representations, postmodern critiques of science argue that representations create reality. Signifying acts of scientists are responsible for generating reality rather than reflecting it. In science, as elsewhere, "we simply cast around for a vocabulary which lets us get what we want" (Rorty 1982, 152). This being the case, the distinction between fiction and nonfiction no longer seems to hold. If the postmodernists are successful, academic divisions, as well as library shelves, will need to be rearranged and dehierarchized. Herein, it appears that the humanities finally begin to take their revenge: The once marginalized become the center.

Notes

1. The sociology of knowledge and the sociology of scientific knowledge have also been important players in this move to examine the noncognitive factors that influence scientific knowledge. I will reserve discussing these movements, however, until chapter 5 since they have a somewhat unique intellectual history.

2. For example, Sandra Harding wishes to build a feminist epistemology on the recognition of the social position of women. Since women are, by the gender division of labor in society, relegated to marginal positions, they are better able to offer radical critiques of science and knowledge. Marginalizing allows insight not available to those in positions of power. A feminist empiricism that simply utilizes male techniques of knowing is insufficient for developing a feminist epistemology. Code (1991, 323) argues that the quest for objective knowledge be "balanced against ecological and emancipatory projects." Duran contends that a feminist epistemology is possible if one avoids normative accounts and instead relies on a naturalized approach to knowledge.

3. Gross and Levitt are also concerned with the views of science expressed by sociologists of scientific knowledge. They reserve a great deal of their venom for Bruno Latour and his actor-network theory of science. There are also other venues where the postmodern approach to knowledge has come under scrutiny: a 1994 conference on science sponsored by the University of Durham (the proceedings were later published in the *Times Education Supplement*) and a 1995 conference sponsored by the New York Academy of Science with the alarmist title, "The Flight from Science and Reason."

Chapter 4

Society, Knowledge, and the Legacies of Durkheim

In the previous three chapters, I first examined the boundary divisions and conventions of the modern episteme and how these demarcations allowed for the development and growth of the doctrine of scientific realism. Next, I explored some of the postmodern attempts to deconstruct those conventions, beginning first with the conceptual divisions and hierarchies of philosophy and literature and then extending into the divisions and hierarchies of all forms of knowledge, including science. So, to this point I have examined two approaches to knowledge—what can conveniently and simplistically be labeled the modern and the postmodern. In this chapter, I turn toward a third approach to knowledge—accounts found in the discipline of sociology, particularly the branch of sociology known generally as the sociology of knowledge.[1] Like philosophy, linguistics, and rhetoric, the sociology of knowledge also has a long history of wrestling with the issues of truth, reality, and knowledge. It too has tried to develop strategies to account for the immanent and transcendent character of modern knowledge and the contradictions this characteristic creates.

The strategies found in the sociology of knowledge to account for knowledge parallel in many ways those found in philosophy and epistemology, but they do have a somewhat unique historical and intellectual trajectory. In many instances, sociological accounts of knowledge have tried to maintain a distance from those found in philosophy and epistemology. In some cases, they have gone so far as to argue that traditional philosophical versions of knowledge are counterproductive and self-defeating. Often the sociology of knowledge has tried to chart a course for understanding knowledge that is grounded in a "social reality" as opposed to a transcendental or natural reality. It has tried to show how knowledge always bears the mark of society.

Sociology, and the social sciences in general, prove to be an interesting case for an examination of the larger debate between scientific realists and postmodern textualists. This is so because sociology is awkwardly situated at the intersection of the boundary conventions of the modern episteme. In other words, the discipline of sociology has been caught between the desire to offer claims grounded in scientific methodology and the reflexive recognition that all knowledge, even the scientific and sociological, is a product of society and culture. Caught between the epistemic poles of scientific realism and humanistic relativism, sociology has a long history of trying to sort out its object of study and methodology. Sociology is a clear example of a discipline caught between the modern realist desire to have firm epistemic foundations and the relativists' antirealist reminder that all such attempts are deeply flawed.

The answer to the epistemic position of sociology has been the paradoxical doctrine of social realism. While there are many different manifestations of social realism, all forms share the conviction that all or most of reality is, in some manner, social in origin. This may include the argument that all reality is socially negotiated via symbolic interaction or the conviction that the rules, norms, or economic structure of society determine reality orders. Most forms of social realism also agree that it is possible to study and explicate social reality by using standard empirical methods of inquiry, such as field work, surveys, or experiments. In this chapter, I explore the development of the doctrine of social realism as it was presented in the work of Emile Durkheim. I survey the various epistemological and substantive impulses and dilemmas left by his legacy for the sociological examination of knowledge. The reason for selecting Durkheim over other classical social theorists, such as Karl Marx or Max Weber, is that it is in his writings where we arguably find the most concerted attempt to "reconcile a humanistic concern for the meanings of social facts with a desire to follow the methods of the natural sciences" (Schmaus 1994, 4). His writings reveal one of the most sophisticated, yet contradictory, attempts to allow for the human impact on knowledge, while still preserving the possibility of a truly objective knowledge of society.

It is worth noting at the outset that in Anglo-American sociology, primarily under the influence of Talcott Parsons and quantitatively oriented sociologists,

Durkheim's writings are often considered to be inspiration on how to make sociology as scientific as the natural sciences. There is certainly enough evidence to support such a reading, particularly if one focuses only on *The Rules of Sociological Method* (1982a) or *Suicide* (1951). In these works, Durkheim proclaimed sociology's intellectual mission as the search for objective and external "social facts." Durkheim's writings, however, particularly his more "social anthropological" writings, conveyed a much more complex epistemological endeavor: to account for all knowledge in sociological terms. The approach he advocated has led to much confusion over his position on the relationship between reality and social representations. His approach has been labeled realist or nominalist (or sometimes both; see Nye and Ashworth 1990) or as positivist or relativist. It has also spawned numerous interpretations and debates regarding his actual final stand on the possibility of obtaining objective knowledge of society (see Gieryn 1982, 108–110; Gane 1988, 103).[2]

In this chapter, I will make no attempt to determine which of these accounts is closer to the real Durkheim. Such a designation is probably meaningless and unimportant for the project at hand. What I argue here is that Durkheim introduced three theoretical and empirical impulses into sociological accounts of knowledge. One impulse, resulting from his epistemological writings, sought to pave the way for the use of scientific methods in sociology. This required a strategy that, in some manner, made scientific methodology a special and privileged way of knowing. It is this strategy that allowed for the epistemological development of the doctrine of social realism. The second impulse, resulting from his substantive theory of culture, attempted to explain all conceptual elements of society or collective representations with reference to varying forms of social morphology (see Gieryn 1982). This second impulse called for cognitive structures, particularly religious and "primitive" ones, to be treated as collective cultural representations that result from the organization of society. This impulse was responsible for the substantive development of the doctrine of social realism. This social realism eventually led, however, to a general reflection on the nature of all knowledge, including the scientific and social scientific. It is this third, more reflexive impulse that eventually called into question the entire epistemological project associated with social realism and sociological explanations of knowledge in general. In simplified terms, the reflexive impulse posited the dilemma: if all knowledge is socially produced imagery, what about sociological knowledge about knowledge? Is it too just a contextually dependent representation or, more precisely, a rerepresentation of the representations of others? In turn, this question also raised still other questions about the relationship between society and scientific knowledge. It is at this point that Durkheim's social realism introduced what could be called a "spell of representation" into the sociology of knowledge (see Pickering 1994). This spell involved a devoted attempt to reconcile scientific objectivity with the idea that all knowledge was a product of socially constituted representations.[3] As we shall see in the next chapter, this dilemma has had a lasting impact on the various theories found in the twentieth-century sociology of knowledge. In what follows, I want to consider the manifestations of these different impulses and dilemmas in Durkheim's writings.

The Epistemological Roots of Durkheim's Social Realism

Durkheim attempted to situate his sociological epistemology in opposition to and at the intersection of both empiricism, which argues that categories of thought are derived from experience, and an a priori rationalism, which argues that categories are universals given by the natural organization of the human mind (see Lehmann 1993, 119–158).[4] In distinguishing his position from previous forms of rationalism and empiricism, Durkheim (1965, 31–32n) wrote:

> The rationalism which is imminent in the sociological theory of knowledge is thus midway between the classical empiricism and a priorism. For the first, the categories are purely artificial constructions; for the second, on the contrary, they are given by nature; for us, they are in a sense a work of art, but of an art which imitates nature with a perfection capable of increasing unlimitedly.

Durkheim's position was one that retained the basic vision of rationalism: to logically capture the reality that exists outside of knowledge about that reality. Unlike previous forms of rationalism, however, which argued that categories of thought were derived from a naturalistic predisposition of the mind, Durkheim saw these categories as given by the social organization of society. For Durkheim, society contained both the limitations and possibilities of rational thought. As societies evolved from primitive to complex, the socially derived categories became more perfected and, as a result, more in concurrence with reality. Consequently, Durkheim believed that it was only in modern complex or functionally differentiated societies where the rational and objective knowledge of science was even a possibility. It was only here that society was capable of better apprehending reality.

Durkheim used this social epistemology to argue that sociology should be removed from the historical influences of philosophy. In his view, philosophy "could only hinder it [sociology] from growing into a positive science" (Durkheim 1982b, 237). Durkheim also argued that reformulating philosophical problems and questions in sociological terms would allow these questions and problems to "at last be liberated from the tautologies in which they have languished" (Durkheim and Mauss 1963, 88). In Durkheim's view, there was no need for sociology to become bogged down in the "grand hypotheses which divide the metaphysicians" (Durkheim 1982a, 159). It must remove those grand hypotheses and metaphysical conjectures from consideration in order to get on with the business of building a positive science of society. In the end, however, Durkheim felt that sociology's impact would prove revolutionary to philosophy. It would provide a means for overcoming a host of philosophical dualisms, such as mind/body, idea/material, and nominal/real, which had long haunted and occupied Western thought. In fact, Durkheim put this in stronger terms when he argued that philosophy must come to rely on the positive sciences or become just another "form of literature" (Durkheim 1982b, 237).

Durkheim's social realism can be seen in its most developed form in *The Rules of Sociological Method* (1982a). In this work, Durkheim (1982a, 33) asserted that the task of sociology was "to extend the scope of scientific rationalism to cover human behavior." He declared that the discipline of sociology should be concerned with the study of "social facts." Social facts were defined as "manners of acting, thinking and feeling external to the individual, which are invested with a coercive power by virtue of which they exercise control over him" (Durkheim 1982a, 52). Social facts were external and collective forces that acted on and within the individual, regardless of the individual's awareness of their presence. Since social facts were to be seen as external to the individual, they should be treated as things, much like nature was treated by natural scientists (see Durkheim 1982a, 69). Like nature, social objects were to be dealt with as "natural phenomena, subject to necessary laws" (Durkheim 1982d, 195). The accomplishment of the sociological examination of social facts required that sociologists, like natural scientists, be both objective and guided by the rigorous controls of methodological procedure. Sociologists must reject information that is drawn exclusively from personal sense experience and instead rely on data that is objective and verifiable (see Durkheim 1982a, 81). The sociologist must "take on the state of mind of the physicist, chemists, and biologists" (Durkheim 1982c, 246). Echoing Bacon, Durkheim (1982a, 72) argued that the mind has a "natural disposition" to repeat past errors unless it is disciplined by a scientific method" (see also Giddens, 1978, 244–245). The scientific method was, thus, the only reliable means of removing the subjective and mythological material that inhibited the apprehension of the social world (see Gieryn 1982, 120). Sociological method could, if it emancipated itself from the "fallacious ideas that dominate the mind of the layman," develop a unbiased and accurate picture of the all-encompassing impact of social facts on the individual and everyday life (Durkheim 1938, 32).

As far as science was concerned, Durkheim's epistemological writings showed a strong proclivity to treat science as a special way of knowing that allows for a "direct contact with things" (Durkheim 1965, 270). In this sense, the modern way of knowing was different from and preferable to that found in primitive societies. Durkheim argued that modern scientific classifications, unlike primitive, lay, or religious ones, were a reflection of the actual nature of things. For example, in describing the urgent need for a science of ethics, he claimed that it is the structural rules of ethics, rather than our conceptualization of them, that "are actually the subject matter of science, just as actual physical bodies, and not the layman's idea of them, constitute the subject matter of physics" (Durkheim 1938, 23). In this sense, scientific classification contained "autonomous realities" that were beyond the influence of the myths and biases of society or lay opinion (Durkheim 1953, 31). Science was a special way of knowing because it was an outcome of a unique and functionally diverse form of society.

Durkheim (1982a, 33) felt that his scientific approach to studying society would be well received by all who "share our faith in the future of reason." In a time of "resurgent mysticism" and unending metaphysical debates, the discipline of sociology, firmly grounded in the doctrine of social realism and the methods of science, was to provide a new means for extending objectivity and rationality into new domains of inquiry. Sociology held the possibility of forever solving the metaphysical problems that had lingered for so long without resolution in philosophy. In some ways, it was the only hope of staving off irrationality and nihilism. Durkheim believed, however, that salvaging rationality required a moral commitment to truth. "The value which we attribute to science depends upon the idea which we collectively form of its nature and role in life"

(Durkheim 1965, 486). It was imperative that not only sociology but also all of modern society be committed to the goals and methods of science or what he referred to as the "cult of truth." This was necessary in order to rescue Western culture from anomie and the return of mystical irrationality.

Durkheim's Substantive Theory of Knowledge

When taken in isolation from his other arguments, Durkheim's epistemological writings reveal a person deeply committed to the ethos and methods of science. In this commitment, he attempted to provide a clear and coherent vision of sociology's object of analysis, a description of what the discipline should be doing, and how its knowledge differs from other forms. If we consult only Durkheim's epistemological writings, however, we would miss his important contributions to the sociology of knowledge and culture, as well as the subsequent paradoxes he posed for the sociological examination of knowledge.

It was perhaps in Durkheim's writings on the relationship between collective representations and social organization (Durkheim and Mauss 1963; Durkheim 1965; 1983) where his sociology of knowledge and his substantive social realism became most fully developed. As we shall see, it was also the point at which the problematic relationship between knowledge and society began to be introduced into the sociology of knowledge. In Durkheim's sociology of knowledge, knowledge was primarily a collective representation explainable by external social factors or social facts. These social facts arose from varying forms of social organization and moral sentiments present in various groups. In an often quoted statement from *Primitive Classification*, Durkheim and Mauss (1963, 82) argued that "it is because men were grouped, that in their ideas they grouped other things." This strong form of social realism argued that all "logical relations are, in a sense, domestic relations" (Durkheim and Mauss 1963, 84). The relationships between objects were a reflection of the relationships between people. Classification schemes and logical systems reflected the type and rigidity of the associational relations among members of a particular collectivity, rather than being accurate or inaccurate representations of the universal or the transcendental logic of nature or God. Mental states of the group depended on the way the group is "founded and organized, upon its morphology, upon its religious, moral and economic institutions" (Durkheim 1965, 28). Even abstract and amorphous concepts such as space and time "are, at each point in their history, closely connected with the corresponding social organization" (Durkheim and Mauss 1963, 88).

For Durkheim, classification activities reproduced the internal and external social boundaries of a group (see also Bloor 1982, 267; Douglas 1982; Fuchs 1992). Every form of classification "is the ensemble of mental habits by virtue of which we conceive things and facts in the form of coordinated or hierarchized groups" (Durkheim and Mauss 1963, 88). For instance, societies with a rigid sense of solidarity tended to have firm boundaries between individual elements of a larger classification system. These cohesive or ritually dense societies typically had elaborate and rigid symbolic and hierarchical differentiations between true and false, right and wrong, bird and fish, space and time, workers and rulers, etc., as well as elaborate rituals to enhance these socially constructed divisions. In such settings, "things are above all sacred or profane, pure or impure, friends or enemies, favorable or unfavorable" (Durkheim and Mauss 1963, 86). Under these circumstances the social and moral order of the group was raised to the transcendental level of absolute, nonnegotiable truth: the mundane was made holy.

For Durkheim, representations were not only reflections of a social order, they were also constitutive elements of it. In this respect, shared categories were necessary for the existence of society. In fact, in many respects they were society. For a society to exist it must have shared conceptions of things such as time, cause, number, and space. Without these shared conceptualization all social life would be impossible (see Durkheim 1965, 490–491). Shared representations, whether of religious or secular origin, were the material that both created the collectivity and holds its members in place. Furthermore, these social conceptions and categories of human thought were "never fixed in any one form; they are made, unmade and remade incessantly; they change with places and times" (Durkheim 1965, 28). Since conceptions and categories were always changing, Durkheim (1965, 25) concluded that they must "depend on factors that are historical and consequently social." But, was this view of knowledge as a collective representation also applicable to the natural and social sciences? Durkheim revealed his more nominalist and relativist side when he showed a tendency in his substantive writings on social morphology to treat all representations as symmetrical, including scientific and lay and modern and primitive ones. For example, Durkheim and Mauss (1963, 88) wrote:

> the history of scientific classifications is the history of stages by which this element of social affectivity has progressively weakened, leaving more and more room for reflective thought of individuals. But it is not the case that these remote influences which we have just studied have ceased to be felt today. They have left behind an effect which survives and which is always present...

From this view, there was no great logical gap between the modern scientific and premodern mythical ways of thinking and categorizing. Both science and myth bore the stamp of collective organization on their knowledge. "Primitive classifications are not therefore singular or exception, having no analogy with those employed by more civilized peoples; on the contrary, they seem to be connected, with no break in continuity, to the first scientific classifications" (Durkheim and Mauss 1963, 81). Consequently, all scientific ideas, such as the conceptualizations of force and gravity, were ultimately of religious origin (see Durkheim 1965, 234). Scientific symbols and classification schemes, like lay ones, "are never more than approximative" (Durkheim 1965, 486). In this sense, belief in science was not fundamentally different than religious faith (see Durkheim 1965, 486). Both required a commitment to something that was recognized as being of a higher order and of transcending the individual.

These writings on the relationship between social morphology and knowledge begin to reveal the reflexive paradox involved in explaining knowledge in sociological terms. The paradox involves, at least in part, Durkheim's attempt to view representations as reflections of social organization while still managing to safeguard his own representations as scientific and truthful. In Durkheim's epistemological writings he argued that science is an objective and rational way of knowing brought into being by the unique social form associated with modernity. In his writings on religion, morality, and knowledge, however, he contended that "in social life, everything consists of representations, ideas and sentiments..." (Durkheim 1982c, 247). This "everything" would arguably include society itself, as well as both scientific and social scientific knowledge. This was so because scientific symbols, concepts, and classifications were products of the confines of particular groups and, consequently, could be seen as part of the "vast symbolism" of a group's social order (see Durkheim 1965, 264). Here, all of "collective life is only made of representations" (Durkheim 1973, 16). In sociological terms, nothing could differentiate the quality of scientific and lay conceptual reasoning. Consequently, science, like all forms of knowing, "is an eminently social matter" (Durkheim 1980, 121). Science was social because it involved "a vast cooperative effort; because it presupposes certain methods and techniques that are the creation of an authoritative tradition," and because "it brings into play ideas that dominate all thought and seemly condense the whole civilization" (Durkheim 1980, 121).

So, on the one hand, Durkheim recognized that all representations were constructed like a "work of art." On the other hand, however, he argued, like many of his predecessors in natural philosophy, that some categories (e.g., scientific and social scientific ones) were capable of evolving toward an ultimate mirroring of reality. This essentially is the crux of the dilemma that Durkheim's social realism found itself in: knowledge proceeded through stages of development ending in science, but these stages were crafted, context-dependent representations constructed by society. They were in the end stories or origin myths constructed by society. As such they were not vastly different from the stories told in a "primitive" society.

Durkheim's Encounter with Pragmatism and the Problem of Reflexivity

Further elaboration and defense of Durkheim's approach to knowledge and science can be found in the "Pragmatism Lectures" (1983). It was arguably in these lectures where the tension between Durkheim's social realism and his reflexive impulse became the most pronounced. Marcel Mauss considered these lectures to be "the crowning achievement of Durkheim's philosophical work" (Mauss in Lukes 1973, 485). In these lectures, Durkheim sought to critique the individualistic and psychologistic approach to knowledge found in American pragmatism and to explore in detail "the nature and function of concepts and the role of speculative thought" (Durkheim 1983, xl). Durkheim contended that pragmatism was symptomatic of the intellectual and cultural anomie of modern societies. In his view, pragmatism, as a scholarly movement, was responsible for intensifying modern intellectual anomie, rather than solving it. His primary complaint was that pragmatism used and advocated a series of trial and error techniques to forge truth. If followed, these individualized techniques would soon lead to "disillusionment with science in general" (Mestrovic 1988, 79). Durkheim (1983, 68) argued that for the pragmatists:

> Reason is placed on the same plane as sensitivity; truth on the same plane as sensations and instincts. But men have always recognized in truth something that in certain respects imposes itself on us, something that is independent of the facts of sensitivity and individual impulses...

Essentially, Durkheim wanted to salvage science, including his own science of society, from the anomic, hyperindividualistic, and relativistic theory of truth found in pragmatism. For Durkheim, pragmatism was no solution to the metaphysical problems of modern knowledge, only a intensification of them.

There was, however, an important issue on which Durkheim and the pragmatists were in agreement: the attempt to "soften truth" (Durkheim 1983, 67). Like pragmatism, Durkheim also argued that:

> History begins nowhere and ends nowhere. Everything in man has been made by mankind in the course of time. Consequently, if truth is human, it too is a human product. Sociology applies the same conception to reason. All that constitutes reason, its principles and categories, has been made in the course of history. (Durkheim 1983, 67)

Both Durkheim and the pragmatists had a "heightened sense of human reality, the feeling for the extreme variability of everything human" (Durkheim 1983, 71). The pragmatists, however, took this insight about the construction of truth to mean that the only conceptualization of truth that was possible was one where it was conceived as a convenient instrument for the attainment of individual goals and wishes. For the pragmatists, truth had only a practical or utilitarian value. Truth was nothing more than a helpful means for organizing individual experience and a useful instrument for the conduct of everyday life. This being the case, pragmatists concluded that culture-at-large had no real need for any broader, cumulative, or speculative philosophy: Knowledge work was to be left entirely to individuals.

It was on this point where Durkheim and the pragmatists parted company. In his view, the revelations of the human character and instability of truth did not mean that it had ceased to serve as a guiding principle for a society or group. For Durkheim, the central epistemological problem was that the pragmatists fail "to recognize the duality that exists between the mentality which results from individual experiences and that which results from collective experiences"

(Durkheim 1983, 68). Durkheim argued that truth, like morality, had an "obligatory nature," rather than merely a philosophical or epistemological one (Durkheim 1983, 98). In his words: "We are not free in a state of certainty. We feel obliged to adhere to truth. We see our certainty as something that is not personal to us, and that is to be shared by all men. Whether this is an illusion or not, we have that belief" (Durkheim 1983: 101). Truth, then, was something beyond individual experience and utility. It could not be adequately understood in the subjectivistic and utilitarian terms used by the pragmatists. Truth needed to be sociologized in order to be properly understood. This was so because only sociology was capable of reminding us that "what is social always possesses a higher dignity than what is individual" (Durkheim 1983, 68). Sociology was the only science capable of analyzing "even that august thing, truth" (Durkheim 1983, 68).

So, Durkheim agreed with the pragmatist's contention that truth had no ultimate metaphysical foundation and that we, therefore, could "no longer accept a single, invariable system of categories or intellectual frameworks" (Durkheim 1983, 71). The sacred character of truth did, however, serve a purpose: It was a means to orient collective existence and direct the social production of knowledge.

> It could well be that certainty is essentially something collective. We are certain that we are not the only ones who are certain. Even when we have worked out a personal belief, we need to communicate it, in order to be certain that we are not mistaken. The authority of tradition and opinion is not, of course, exempt from criticism. When we criticize them, however, it is always in their own name. When, for example, we criticize popular prejudices in the name of science, we are using the authority which opinion accords to science. (Durkheim 1983-102)

Both pragmatism and Durkheim's sociology of knowledge concurred that certainty and truth were not absolutes derived from reality. Unlike the subjectivistic implications drawn by pragmatism, however, Durkheim emphasized that certainty and truth were socially contingent issues that were worked out and negotiated through collective experience. As in everyday life, if there were disagreements on truth claims, there was an appeal to legitimate authority structures to determine the outcome of competing claims.

As in his writings on collective representations and social morphology, such a sociological view of knowledge also seems applicable to science. In these lectures, Durkheim (1983, 105) himself argued that "the concepts worked out by the masses and those worked out by scientists are not essentially different in nature." He wrote:

> Concepts which are collective in origin (as all concepts really are) take on in our eyes, even when their object is not a real one, such a strength that it appears to be real. That is why concepts acquire the vividness and force of action, of sensations. (Durkheim 1983: 101)

Yet, in this setting of the arbitrary nature of representations, conceptual reasoning, like social facts, retained a social reality. Sociologically speaking, it was unimportant that the symbols did not mirror reality, for all reality was always and everywhere socially constructed.

As in his social epistemology of the *Rules*, one of the key issues Durkheim raised in these lectures was that truth and reality could not be understood adequately using traditional philosophical accounts. This was so because truth was a socially derived concept, not an epistemological one. In fact, there was really no correspondence between the social version and nature of truth and the philosophical version. Within the framework of traditional epistemology, it may indeed be impossible to establish firmly a strategy for uncovering absolute truth. Sociologically speaking, however, this does not matter. Socially the ideal of truth still has meaning. Concepts such as truth and reality have a certain social facticity beyond the metaphysical debates found in philosophical discourse. Hence, the collapse of meaning in philosophy does not tell us anything about the collapse or noncollapse of the social organization and cultural production of reality and truth. It tells us only why one particular productional rationale of truth and one version of reality were no longer possible.

The crisis of turn of the century epistemology, represented most dramatically by Nietzsche and pragmatism, had cast doubt on all attempts at establishing correspondence between word and world. Durkheim (1983, 68), however, argued that "it is one thing to cast doubt on the correspondence between symbols and reality: but it is quite another to reject the thing symbolized along with symbol." The "world is always filtered through our ideas of it...that fact in no way negates the 'objectivity' of the world" (Meštrović 1988, 24). Indeed, the relationship between signifier and signified may be arbitrary and a correspondence between world and word may not exist, but this did not mean for Durkheim that these distinctions are socially meaningless or cease to be of social importance. "The pressure that truth is seen as exercising on minds is itself a symbol that must be interpreted, even if we refuse to make of truth something absolute and extra-human" (Durkheim 1983, 68).

Durkheim's pragmatism lectures clearly illustrate his recognition of the historical variability of truth, as well as its constructionist and nonrepresentational character. Durkheim, however, unlike other turn of the century writers such as Nietzsche and James, did not completely abandon the concept of truth or psychologize it. He used these lectures to resituate truth as the symbolic manifestation or collective representation of the moral sentiments and organizational dynamics of a particular collectivity. For Durkheim, "truth is a norm for thought in the same way that the moral ideal is a norm for conduct" (Durkheim 1983, 98). Like religion, truth served to united people into a moral community. It may be philosophically relative, yet it was socially real and confining.

Conclusion: The Legacies of Durkheim

The point of examining Durkheim's different epistemological and substantive writings is not to reveal that he contradicted himself, since most theorists do at some point, but to demonstrate how he introduced three central epistemological and theoretical impulses into the sociological examination of knowledge. These three impulses resulted in two divergent realist and relativist readings and messages. In the first realist message, we find a strict separation of scientific and nonscientific ways of knowing. In this message we see what Fleck (1979, 47) referred to as "an excessive respect, bordering on pious reverence for scientific facts." This social epistemology was extended into the substantive study of knowledge where a firm division is drawn between the external cause of knowledge, society, and the internal effect, collective representations. In the second, more relativist, message it became unclear as to where to draw the line between society and knowledge or if drawing such a line was even possible, since everything could be viewed as a collective representation, including science, social science, and even the knowledge about collective representations. In the first message, it was argued that on the one side are "effects" such as collective representations, classification schemes, religious knowledge, and belief systems, while on the other were "causes" such as social organization, moral orders, social cohesion, and later structuralism, language and mythemes. Within this message, the social side of the axis was separated and given priority over the ideational.

Knowledge types and styles were explained by reference to some external social condition or sets of social facts beyond the content of the ideas themselves. This sorting of the social from the representational became the task of much of twentieth-century structuralism.

In the second, relativist, impulse however, it became difficult to demarcate society from knowledge because society itself was seen as made up of "a complex of ideas and sentiments" (Durkheim 1961, 277). Here, there appeared to be no external, outside force to cause knowledge, for that force too was but another representation. Within this impulse everything, including society, scientific knowledge, and Durkheim's sociology of knowledge itself, became a representation. Within this scenario, instead of having society on the one side and knowledge on other, there appears only to be representations or what the postmodernists now like to call "texts." In this case, it appears that the "society-as-God" of social realism might simply be the "genie malign" of social relativism (see Godlove 1986, 400). From this reading it seems that the social foundation of Durkheim's social epistemology itself was just another representation. Illustrating how all the world is a representation became the task of late twentieth-century deconstructionists and reflexive sociologists.

It can be argued that these contradictory impulses or messages harbor both "the spell of representation" (Pickering 1994) and the reflexive dilemma of twentieth-century social realistic accounts of knowledge. In the social realist account, the task of the sociology of knowledge became to show how sociological representations were closer to the real or more special than the representations it described. If knowledge was a representation, however, and context dependent and socially determined, why would that not be applicable to the sociological knowledge about the social character of knowledge? Nothing within the substantive account of social realism would exempt sociological knowledge from social influences, nor was there any epistemic protection from the claim that the "social as cause" of sociology was but a representation as well—in this case, the representation of one professional knowledge-producing group. In this case, there could be no absolute standard of truth, not even the seemingly firm "social structure" of sociology and structuralism.[5] Within this reflexive path, Durkheim's social realism was easily capable of slipping into an all-encompassing cultural relativism.

In the relativist account, society as an entity or first principle disappeared, to be replaced solely by representations. Here, the task became to transform everything into an artlike representation. The problem of reflexivity, however, haunted this line of thought as well. As with postmodernism, if all that exists were representations, including the knowledge that all that exists were representations, how is it possible to defend any type of interpretative claim (see Fuchs and Ward 1994a/b)? On what basis were some claims to be elevated over others? Are not the representations of *USA Today* as valid as those of social theorists? Are not the statements of creation scientists symmetrical with evolutionists? Since all accounts are by necessity inaccurate social representations of the world, what makes some better than others? In other words, why are some representations preferable over others? What guides the selection of explanations? The second Durkheimian impulses and legacies "asks, but cannot answer, how it is that we are able to represent reality given that our central categories have a social origin" (Godlove 1986, 400).

So what, then, is sociology and what is to be its disciplinary mission? Durkheim (1953, 70) himself had earlier wondered if this "enterprise" of sociology was possible or if it was "contradictory in terms." Perhaps it is possible to conclude that Durkheim was able to suspend reflexivistic thinking because of the desire to found and promote an academic discipline. The point of establishing a science was to develop a way of explaining things not necessarily to explain the act of explaining. In the context of the late nineteenth century, explaining the act of explaining was something to be left to philosophers, metaphysicians, and epistemologists. Such a quandary was often viewed as a holdover from the history of philosophy for which there was no room in modern science. Such a concern was not, nor should it have been, a consideration for those developing a truly positive science of society.

The postmodern condition has made such dismissals of reflexivism in the name of building a science much more problematic. Postmodern theory directly calls attention to the nature of all knowledge claims, not just the claims of others. As Malcolm Ashmore (1989, 69) has pointed out, postmodernism destroys rationalism in two nihilistic stages. It first posits a relativism that sees reality and truth as context dependent. It then puts forth a reflexivity that argues that even the contention that reality and truth are contextual is itself context dependent. In doing so, postmodernists want to force all forms of knowledge, in some way, to encounter the forever-problematic relationship between representation and reality.

In the next chapter, I explore how twentieth-century sociology and social theory has incorporated the impulses of Durkheim and how it has tried to resolve this nagging problem of reflexivity.

Notes

1. The roots of the sociology of knowledge have at least three overlapping branches. One branch can be linked with Marx and his concern for the relationship between the material organization of society and its superstructural belief systems. Another branch leads from Weber's comparative studies of religion and culture. A final branch, and the one to be considered in detail here, begins with Durkheim's attempts to show how the social organization of society influences social representations.

2. Some commentators, such as Steven Lukes (1982) and Mary Douglas (1975), see Durkheim's work as never successfully resolving the tension between socially produced representations and scientific objectivity. From this viewpoint, Durkheim's positivism gets in the way of his larger project to explain knowledge. Others, such as Warren Schmaus (1994), see his work as carefully and successfully negotiating a path between social realism and cultural relativism, while still others, such as Stjepan Mestrovic (1988, 1991), argue that Durkheim's approach can best be thought of as a type of reconstituted rationalism that cannot be easily placed in any previous or existing "ism."

3. The problematic relationship between knowledge and society undoubtedly has a number of other sources, such as Marx's discussion of ideology, the neo-Kantian methodological approach of Weber, and Schutz's and Garfinkel's phenomenological and ethnomethodological approaches to knowledge.

4. Durkheim's sociology of knowledge and Nietzsche's equation of knowledge with power can be seen as two of the most influential late nineteenth-and early twentieth-century attempts to replace philosophical versions of the origins of knowledge, particularly empirical and a priori ones, with other, less philosophical versions. Despite important differences between Nietzsche and Durkheim, both attempted to understand knowledge with reference to social conditions and processes rather than to preexisting transcendental categories or inherent empirical possibilities (see Ward 1994).

5. Within the reflexive irony that Durkheim's social realistic dualism eventually created, humans came to be understood as "speaking myth," but, at least in the case of Levi-Strauss (1969), were somehow able to speak truth when speaking about myth.

Chapter 5

Dealing with the Durkheimian Dilemmas

For the sociology of knowledge, Durkheim introduces, or at least intensifies, a set of plaguing problems associated with representation, knowledge, and society. In general terms, these problems involve explaining the relationship between knowledge and society and, more problematically, defending this sociological knowledge about knowledge—coming as it does from context dependent and historically and socially situated actors. The first step is relatively straightforward and, when taken alone, seemingly less problematic. Sociology simply uses its standard theoretical and empirical repertoire to show how class, gender, culture, social organization, interests, or some other "social fact" affects, distorts, or impedes the production, translation, or reception of knowledge. The real epistemological quandary comes when sociologists try to defend their truth claims about truth from charges of social reductionism or when they are forced to become reflexive and turn sociological analysis on their own knowledge about knowledge. In the first case, sociologists face the task of trying to show how their truth claims have more validity or are truer representations of reality than the indigenous claims of those who they study or the claims found in competing knowledge fields such as psychology or biology. For example, what makes sociological declarations about the conduct of science closer to the reality of scientific activity when compared with scientists' own accounts? In the second case, when sociologist are forced to become reflexive and turn their own methods and perspectives on themselves, their social realism becomes even more problematic. For instance, are sociological accounts merely secondary representations of the representations of others? Are not sociological representations also situated in society and, therefore, culturally relative? If social forces affect all knowledge, then the knowledge that social forces affect knowledge must, too, be a product of society. In this case, it quickly becomes unclear why sociologists should try to play the representation game anymore since all representations are fictitious constructions.

In this chapter, I consider a few of the sociologies of knowledge produced in the twentieth century. Specifically, I examine how these sociologies of knowledge have incorporated the Durkheimian impulses and how they have sought to deal with the representational dilemmas left by him. All of the approaches to be considered in this chapter try to explain knowledge sociologically (broadly defined) and all in some manner must confront the epistemological status of their own sociological knowledge about knowledge. In the following, I look at four general sociologies of knowledge. First, I examine accounts that follow the first Durkheimian impulse and exempt scientific and sociological knowledge from social critique. These accounts can be found in widely varying forms in the writings of Karl Mannheim and Robert K. Merton. Second, I explore the more microsociology of knowledge found in phenomenology, ethnomethodology, and Habermasian sociology. These accounts have a somewhat different intellectual origin than the others examined in this chapter; however, they too are concerned with explaining the social dimensions of knowledge and defending this knowledge about knowledge. Third, I examine the full-blown substantive social realism found in the strong programme, the sociology of scientific knowledge (SSK), and social constructionism. These approaches follow the second impulse found in Durkheim. They attempt to complete his application of social realism to religion by providing an analysis of all forms of knowledge. Finally, I briefly survey what is called the reflexivity project and its experimentation with new literary forms or modes of (re)presentation. This account emerges from the third, more reflexive impulse found in Durkheim's writings.[1]

The movements discussed here certainly do not constitute an exhaustive or mutually exclusive list of twentieth-century sociologies of knowledge. In fact, many of these responses feed off and are influenced by one another. In addition, not all of these responses are direct outcomes of Durkheim's writings or the Durkheimian impulses. What I hope to show, however, is that all of these twentieth-century sociologies of knowledge are not adequate for overcoming the general paradoxes left by Durkheim's legacy, nor are they theoretically equipped to handle the theoretical challenges currently put forward by postmodern theory.

Mannheim and Other Free Floaters

The sociology of knowledge was formally introduced as a branch of sociological analysis not by Durkheim but by Max Scheler and Karl Mannheim in the 1920s. Mannheim's intellectual debts were much more Marxist and Weberian than Durkheimian; however, he shared with Durkheim the general attempt to explain culture and knowledge in social and contextual terms. The analytical strategy Mannheim worked out, like Durkheim's, also has come to be criticized as being overly reductionistic and paradoxical.

Mannheim (1936, 1) argued that the goal of the sociology of knowledge was to explore how thinking functions as "an instrument of collective action." In this respect, as with Durkheim, the goal of the sociology of knowledge was vastly different from that of traditional philosophy and epistemology. Philosophers had typically only concerned themselves with forms of pure knowledge and normative means of obtaining it. They had not generally been concerned with the dynamics of everyday commonsensical thought and knowledge. Mannheim (1936, 1) wrote:

> Philosophers have too long concerned themselves with their own thinking. When they wrote of thought, they had in mind their own history, the history of philosophy. This type of thinking is applicable only under quite special circumstances, and what can learned by analyzing it is not directly transferable to other spheres of life.

In Mannheim's view, traditional philosophy, with its quest for absolute and universal truth, was not conceptually equipped to address the historical, cultural, and group variability of most knowledge forms. Philosophers were accustomed to sorting knowledge into the distinct categories of belief or truth through the use of logic or the assessment of empirical findings. They then used this logic or empirical assessment to provide normative bases for all of knowledge acquisition. The sociology of knowledge, however, as Mannheim (1936, 2) conceived it, was to trace the social origins of certain "modes of thought." The task of the sociology of knowledge was to locate knowledge and these modes of thought within the historically and culturally specific social milieux that formed them and the social frameworks that produced them.

In this quest Mannheim recognized that the sociology of knowledge itself was also historically and socially situated. In his view, in a setting where truth can be clearly ascertained there was really no need for a sociology of knowledge. When religion, philosophy, science, or some other discourse could provide the appropriate rationale for the accumulation of knowledge, there was no need for a discussion of the social circumstances that produce it. He wrote:

> It is clear that such problems become general only in an age in which disagreement is more conspicuous than agreement. One turns from the direct observation of things to the consideration of ways of thinking only when the possibility of direct and continuous elaboration of concepts concerning things and situations has collapsed in the face of fundamentally divergent definitions. (Mannheim 1936, 6)

In a historical period where traditional accounts of truth slip away or become dismantled, however, such as in modern society, the sociology of knowledge was seen as a necessity for determining the cause of knowledge. The modern multiplicity of definitions and the resulting scattering of meaning had forced theories of knowledge to evaluate the underlying factors that influence knowledge. In Mannheim's view, this multiplicity and fragmentation had introduced the appropriate time for the emergence of the sociology of knowledge.

It is in this regard that Mannheim's sociology of knowledge can be seen, at least in part, as an attempt to reconstruct the basic rationalistic goals of scientific investigation.[2] Mannheim, like Weber in *The Methodology of the Social Sciences* (1949), wanted to disclose the hidden social and historical dimensions at work in all scientific endeavors in order to provide a limited autonomy to science. Mannheim (1936, 189) argued that the sociology of knowledge will allow us to "calculate more precisely modes of thought and to predict ideological reactions." It would be a crucial instrument for clarifying the factors that inhibit and restrict rational knowledge of the world.

The intrinsic logic of such a sociological approach would, at some point, have to dispense with the idea that truth claims are evaluated on their ability to correspond with the object of analysis or the intellectual integrity of what is claimed as true. In other words, the sociology of knowledge would have to suspend the epistemological quest of determining who is telling the truth, since it possesses sociological rather than exclusively logical or empirical criteria of judgments. Because Mannheim viewed the sociology of knowledge as partly a reconstructivist project (i.e., a means of determining the factors that distort truth), he was reluctant to open all forms of knowledge to sociological examination.

Essentially, he relegated the focus of the sociology of knowledge to what might be called mundane discourse or nonserious speech. The latent rationalism in Mannheim's approach forced him to treat science, logic, mathematics, and sociology as special cases of knowledge. These fields and their practitioners were seen as "free floating." They were not bounded by their cultural context, nor were their knowledge products.[3] The freefloating sciences were, therefore, not potential sites for the sociology of knowledge (see Mannheim 1936, 79).

Another strategy found in twentieth-century sociology to make the natural and social sciences a special and privileged way of knowing can be found in the writings of Robert K. Merton (1973). Unlike Mannheim, Merton sought to safeguard rationality by grounding knowledge in the social organization of the institution of science.[4] As with Durkheim's epistemological defense of modern forms of knowledge, Merton's strategy essentially focuses on the unique organizational norms found in science. From his view, science exemplifies the distinct organizational norms of communism, universality, disinterestedness, and organized skepticism (Merton 1973). These norms, when institutionalized into the organization of science, constrained the cognitive and practical activities of scientists. They ensured that all claims were carefully analyzed, weighted, and reviewed before being accepted as true. Consequently, the integrity and truthfulness of science does not rest on the moral characteristics of individual scientists, but on the "rigorous policing" of the institutional mechanisms of science (Merton 1973, 276). Science is truthful because it is a unique and complexly organized institution.

This first tactic for defending sociological knowledge found in the Mannheimian and Mertonian sociology of knowledge seeks to account for the special status of sociological and scientific knowledge by utilizing a strategy of exception and containment. Here, scientific or mathematical statements, including those of sociologists, are to be treated as special cases. They can either be seen as free floating (Mannheim 1936, 79) statements and representations above mundane cognition and inquiry or as containing unique organizational norms not found in other, more ordinary knowledge-producing collectivities (Merton 1973). In either case, scientific truth claims are exempt from the sociological analysis performed on other knowledge claims. This so-called weak program preserves the sacredness of its own proclamations by confining the sociology of knowledge to folk wisdom, lay reasoning, "ideology," and other "wrong-headed" approaches. In this first sociological strategy, accurate representations of social reality are possible because science is a socially special form of inquiry.

Grounding Knowledge in the Lifeworld: Phenomenology, Ethnomethodology, and Habermasian Theory

A second influential view of the relationship between knowledge and society can be found in phenomenology, ethnomethodology, and Habermas's theory of communicative action. These approaches share the conviction that accounts of knowledge should be based upon the dynamics of everyday life rather than a transcendent nature or society. In general, these approaches dispense with official sociological truth claims and epistemological hand wringing in order to examine the everyday reasoning and practices of actors in the lifeworld (see Schutz 1967; Garfinkel 1967; Berger and Luckmann 1967). For phenomenologists and ethnomethodologists, actor accounts are more real and truer than the contrived and second-hand conceptual schemes of traditional sociologists or philosophers. The phenomenological approach heeds the advice found in Berger's and Luckmann's (1967, 13) metaphor that including epistemological problems in the sociology of knowledge is "somewhat like trying to push a bus in which one is riding." While Berger and Luckmann acknowledge the importance of epistemological questions, they argue that they are not a "part of the empirical discipline of sociology" (Berger and Luchmann 1967, 13). Epistemological dilemmas only inhibit the further development of the sociology of knowledge.

In phenomenology and ethnomethodology, the only reality "out there" is that which is constructed, interpreted, and acted on by human actors. For example, ethnomethodology's strategy is to trace and report the movement of actors as they go about the business of managing uncertainty and building reality. In this process, ethnomethodology recognizes an interdependence between appearance and reality (see Woolgar 1988c, 21). Actors move back and forth from object to appearance as they discern and construct meaning. In the end, the meaning that is given to objects or events is a result of an intricate interactional process of social negotiation and interpretation.

In phenomenology and ethnomethodology, the problem of the relativism of sociological knowledge is solved by looking to the specific recipes and practices of people, rather than at the pronouncements of sociologists or other official knowledge producers. This approach dispenses with hypothesis testing and "sociological theory" and instead relies on indigenous "accounts." Ethnomethodology can provide only descriptions and interpretations of reality building, not some allegedly deeper and truer "sociological" reason behind it. In the process of examining everyday life, epistemological problems associated with reflexivity are to be bracketed in order to get on with the description of order-making practices. It is the task of the phenomenological approach to explain how everyday knowledge is created and maintained, not to assess its validity or truthfulness (see Berger and Luchmann 1967, 3). In this sense, ethnomethodologists focus on an "endogenous reflexivity" as opposed to a "referential reflexivity" (Pollner 1991). In other words, ethnomethodology's goal is to examine the reflexivity of actors as they construct meaning in everyday life, not ethnomethodologists as they contemplate their own work about those constructions. In order to preserve naturalistic observation, it is important to analyze "what members do in, to, and about social reality" and not how ethnomethodology itself is "a constitutive process" (Pollner 1991, 372).

Habermas's strategy for explaining knowledge and defending reason and science draws from some of the basic insights of phenomenology and ethnomethodology about the primacy of everyday knowledge. In broad terms, he wants to use phenomenologically based insights to defend reason against irrationality and to provide a normative basis for accessing knowledge claims. Habermas agrees with other postrealist critics who claim that it is no longer possible to establish a foundational metaphysics or first principle to legitimate truth claims. The demise of a correspondence theory of truth and a foundational metaphysics does not mean, however, that we are without any foundations. In the context of everyday life there remain certain rules and structures found in the background of linguistic practices and speech acts that define and legitimate all discursive and normative activities. The rules embedded in communicative action make the quest for truth through consensus a normative activity that guides all speech activities. Habermas (1984, 398) argues that a communicative reason

> refers neither to a subject that preserves itself in relating to objects via representation and action...but to a symbolically structured lifeworld that is constituted in the interpretative accomplishments of its members and only reproduced through communication.

Thus Habermas (1984, 17) concludes that "the rationality inherent in this practice is seen in the fact that a communicatively achieved agreement must be based in the end on reasons."

Communicative rationality serves as a court of appeals for the settlement of truth claims. If the discursive arena is free from the distortions of power, self-interest, and ignorance, then open communication and reason are possible. Under these "ideal speech" conditions, truth claims can be settled by appealing to the background of a particular discursive community. This being the case, truth becomes what is agreed upon under these conditions of ideal speech. Truth is a conversationally dependent and normatively enforced principle. It does not have to be grounded in a first principle or foundational metaphysics. The affirmation of truth is already given in the potentiality of human communication. "Like freedom and reconciliation, truth is built into the linguistic mechanism of the reproduction of the species" (Habermas 1984, 398).

Habermas goes further to argue that scientific knowledge actually approximates an "ideal speech community." Being the closest approximation of ideal speech community that we have, science is able, more than other forms of discourse, to forge truth. In this sense, it is possible for lifeworlds to evolve on the basis of a societal learning process (Habermas 1979). Societies and groups can "be rated and classified according to the extent to which they exemplify the higher forms of social integration and higher problem-solving capacities" (Schmid 1982, 169). Because science involves special forms of communication and interaction, it is a higher, and potentially more truthful, form of knowledge.

While ethnomethodology, phenomenology, and Habermasian theory are coming at the issues of reality and truth in very different ways, they are in agreement that the realm of everyday life holds the key for solving or at least avoiding past epistemological puzzles. All of these movements are convinced that all accounts of knowledge must be grounded in the pragmatics and potentialities of daily life and not the grand, overarching strategies of philosophers or epistemologists if they are to accomplish the task of providing a rationale for truth. Accounts that do not ground knowledge in the lifeworld run the risk of continued abstraction and contradiction.

The Strong Programme and the Sociology of Scientific Knowledge

The Mannheimian paradox found in the so-called weak program in the sociology of knowledge had the latent effect of placing sociological explorations of knowledge in a type of stasis. Within the last twenty years or so, however, the sociology of knowledge has made a comeback in the area of science and technology studies. During this period one of the most empirically active and theoretically innovative subfields within the sociology of knowledge and science studies has been the sociology of scientific knowledge (SSK). Approaches such as Edinburgh's strong programme (see Barnes 1974 and 1982; Bloor 1976 and 1983), the Bath group's Empirical Program of Relativism (see Collins 1985, 1981, 1983; Yearley 1988) and Knorr-Cetina's (1981) social constructivism, to name but a few, have succeeded in ousting the once-dominant Mertonian social studies of science and in reviving the ailing and paradoxical older sociology of knowledge.

Despite important and long-standing differences between these various approaches and camps, they have all attempted to pick up where Mannheim's "nerve failed him" (Bloor 1976, 8). They are also strongly influenced by Durkheim's attempt to establish a causal relationship between knowledge and social organization and morphology. Within these perspectives, following the advice of Mary Hesse (1976, 42), "knowledge is now taken to be what is accepted as such in our culture." In the strong programme no form of knowledge will be treated as free floating, suprahistorical, or special, including scientific and sociological knowledge. All knowledge formations will be treated as epistemologically symmetrical, constructed, and of the world.

For advocates of the strong programme, there is "nothing in the physical world which uniquely determines the conclusions of that community" (Mulkay 1979, 61). What is true for us here and now could, under other circumstances, be completely different. Reality never determines the form in which it will be represented. Rather, that issue is decided by people in practice. From this view, "explanations should be developed within the assumption that the real world does not affect what the scientist believes about it" (Collins and Yearley 1992b, 372). This being the case, all knowledge, including the scientific, is social through and through. Knowledge, like all of social life, is affected by politics, self-interest, and cultural categories. People are not controlled by ideas, but rather through "coercion, the manipulations of rights, or the generation of interests" (Shapin and Barnes 1977, 60). All knowledge is in some manner affected by the social circumstances under which it is produced. In more forceful terms, "selves, minds, and ideas are not merely social products, nor are they merely socially constructed; they are social constructs" (Restivo 1990, 123).

In the seminal work *Knowledge and Social Imagery* (1976, 4–5), Bloor sets out the basic goals and tenets of the strong programme's approach to knowledge:

1. It would be causal, that is, concerned with the conditions that bring about belief or states of knowledge.

2. It would be impartial with respect to truth and falsity, rationality or irrationality, success or failure.

3. It would be symmetrical in its style of explanation. The same types of cause would explain, say, true and false beliefs.

4. It would be reflexive. Like the requirements of symmetry, this is a response to the need to seek general explanation.

The strong programme, like Mannheim's sociology of knowledge, sought to move beyond the discussion of truth and reality found in traditional epistemology. Unlike Mannheim, however, the strong programme treats all knowledge forms as potential sites of sociological investigation. Also, unlike Mannheim, the strong programme is not fearful of the so-called specter of relativism. In fact, relativism is seen by the strong programme as a necessity for doing the sociology of knowledge.

> Our claim is that relativism is essential in all those disciplines which account for the diversity of systems of knowledge...It is those who oppose relativism and who grant certain forms of knowledge a privileged status, who pose the real threat to a scientific understanding of knowledge and cognition (Barnes and Bloor 1982, 22).

For advocates of the strong programme, the fight in the sociology of knowledge and elsewhere over the epistemic position of knowledge is between those who want to grant a privileged position to one form of knowledge over another (rationalists) and those who want to treat all knowledge equitably or symmetrically (relativists).

From the position of the strong programme, it is no longer possible to defend the position that science is hierarchically superior to other forms of knowledge. It is no longer possible simply to brush some knowledge claims aside by arguing that they do not meet the rational standards of scientific respectability. From their view, rationality and irrationality are not "two distinct and qualitatively different classes of things" (Barnes and Bloor 1982, 28). Rationality and irrationality are both explainable utilizing the same sociological approach. Determining what is rational and what is irrational or what is true and what is false is simply a matter of preference based on the prescribed standards existing within the investigator's locality. There is no need to appeal to any higher sources, such as reality, for the determination of those issues because the answer is to be found within the various social contexts that impart the various criteria for judgment. As a result, even knowledge forms that are supposedly empirically and universally valid and reliable, such as science and mathematics, are dependent upon culturally produced categories and systems of meaning. No rationally or empirically based system of knowledge can escape the conceptual boundaries of society.

The strong programme and its relativist approach to knowledge helped launch the sociology of scientific knowledge (SSK)—a branch of sociology concerned with explaining the actual content of scientific knowledge. The strong programme called for sociologists to treat scientific knowledge in the same empirical and realist way that scientists treat nature (see Hull 1988, 4). SSK is designed to explain "what comes to count as scientific knowledge and how is it so to count" (Collins 1983, 267). From the late 1970s onward, it has served as inspiration for a series of empirical studies of scientists at work in their laboratories (see Knorr-Cetina 1981; Latour and Woolgar 1986; Traweek 1988). Operating under the guidance of the empirical program of relativism, these studies explore the ways through which scientists construct and maintain their version of reality and truth. Many of these studies conclude that science is very much a localized craft activity. In this sense, scientists in the laboratory act very much like practical reasoners in everyday life. Scientists, like lay reasoners, utilize the social material at their disposal to construct truth. Essentially, SSK concludes that there are no gaps between lay and scientific reasoning. Both use local resources at their command to make their case stronger while simultaneously weakening the position of others (also see Quine 1964).

In pursuing their analysis of scientific knowledge, sociologists of scientific knowledge make a key distinction between ready-made science and science in the making. Science in the making is rather provisional, ad hoc, highly competitive, and unstable. Here, fact finding is not a rule-governed activity involving the careful application of scientific methodology, as in the accounts of scientific realism. Rather, fact finding is in actuality fact making. It is a constructive process where scientists use the resources at their disposal to build a case that is stronger than that of their opponents (Knorr-Cetina 1981). In science in the making there is often considerable debate over data interpretation and what will stand as an accurate representation of reality. Replication rarely solves the interpretative dilemma either, since it too is fraught with problems of diverging interpretations and methodological critique (see Lakatos and Musgrave 1970, 132–35). On the other hand, ready-made science does appear consensual and rule–governed. Here, it does seem as if truth claims have been carefully evaluated and confirmed. SSK advocates argue that this is not due to a truth claim's accuracy, but because the claim has become "black-boxed." In other words, scientists have so stabilized social order in the laboratory and elsewhere that natural order emerges (see Collins 1985).

Further attention of SSK advocates has been directed at the "inscription devices" used by scientists to produce and report results (Lynch 1985; Lynch and Woolgar 1990; Fyfe and Law 1988). It is here that SSK and rhetorical studies of science most directly connect. Scientific inscription devices do not report unmediated reality but provide a visual representation of something in the cloud chamber or petri dish. These representations serve to produce an "impression of rationality" (Lynch 1985, 1991). They can be seen as objectifying practices that transform the amorphous and local into the concrete and universal. These inscription practices attempt to "convey the impression of precise distinctions, bounded cognitive fields, and delimited causal forces" (Lynch 1991, 13). In SSK, inscription devices become one of the key elements of the overall social construction and social history of truth.

Inspired by SSK's rich empirical analysis of knowledge production and growing out of phenomenology's and ethnomethodology's attempt to explain whatever passes for knowledge in a society has been the far more ambitious and interdisciplinary efforts of social constructionists to extend social realism to every existing object and thing. For social constructionists, everything is a product of society and social interaction and negotiation. Gergen (1985, 266) has defined the focus of social constructionism as examining the "processes by which people come to describe, explain, or otherwise account for the world (including themselves) in which they live." The areas that have seen the most impact of the social constructionist approach are social problems research (see Best 1995) and medical sociology. For instance, in social problems research the concern shifts from looking at social problems as objective entities to looking at them as constructed phenomena. For example, instead of looking at what causes poverty, constructionists examine how poverty becomes constructed as a problem in the first place.

In doing their analyses, social constructionists generally take two lines of inquiry: externalistic and internalistic. Externalist approaches borrow heavily from Marx and Mannheim. They seek to link phenomena with outside, contextual social causes, such as politics, gender distinctions, historical milieus, value systems, and economic changes. On the other hand, internalists borrow heavily from phenomenology and ethnomethodology. They focus more on the issues of negotiation and interaction that make a knowledge claim possible.

In addition to the two lines of substantive analysis in social constructionism, there are also two reactions to its epistemic status. One position has been labeled "procedural," the other, "reflexive" (Manning 1994, 118). Procedural social constructionism wants to get on with the work of describing how things are socially constructed. For proceduralists, like their ethnomethodological predecessors, epistemic problems are a distraction to doing knowledge work. On the other hand, reflexive social constructionists believe that "reflexivity is a revelation about the limits and validity of sociological claims" (Manning 1994, 118). It is, consequently, an opportunity to expand the constructionist approach into new and innovative directions.

The strong programme, SSK, and social constructionism argue that by treating all knowledge as epistemologically symmetrical and socially constructed, including their own, they can overcome the "special case" status of knowledge applied to science, mathematics, and sociology in the Mannheimian sociology of knowledge and the Mertonian sociology of science. These approaches seek to overcome reflexivism through a radical empirical program of relativism that methodically traces and reports the social construction of all truth claims. Advocates of the strong programme and the empirical program of relativism claim that the task of a sociology of knowledge is to study the actual, real conditions of scientists in the laboratory, not to wrestle with epistemological problems. For instance, Collins has urged sociologists to "treat the social world as real, and as something about which we can have sound data, whereas we should treat the natural world as something problematic—a social construct rather than as something real" (Collins 1981: 217). Such "detailed descriptions" are said to dissolve "epistemological mystery and wonder" (Collins and Yearley 1992a,

309). The prescription for remedying reflexivity is to be realist about the observations and findings of sociologists because this is a truly real, causal realm. In other words, social forces are real, but natural ones are socially produced representations.

It is clear that SSK and related programs offer a new approach to understanding knowledge formation. They "simply refuse to play according to the rules and guidelines established by traditional philosophy" (Tibbetts 1986, 53). The central argument of SSK is that all knowledge and all philosophical divisions are socially crafted by knowledge practitioners—they are social through and through (see Pickering 1992). Reality does not have built-in philosophical or scientific distinctions nor secrets waiting to be discovered. Rather, it, too, is a social construct of knowledge practitioners.

The specter that haunted the traditional sociology of knowledge, particularly Durkheim's and Mannheim's, also haunts its contemporary predecessors, SSK and social constructivism. If all statements are epistemologically symmetrical, what makes sociological descriptions of scientific knowledge closer to the reality of scientific activity compared with the scientists' own accounts (see Ashmore 1989)? The principles of symmetry and reflexivity required by the strong programme force SSK to examine its own propositions and social constructions of knowledge. Yet, in doing so, they maintain a constructivist view of science by utilizing a realist view of sociology. "For them Nature is uncertain but Society is not" (Callon 1986, 197). In the zeal to overcome Mannheimian timidity, a reflexive paradox soon reappears, albeit in a somewhat new form.[5]

The Reflexivity Project and New Literary Forms

A more recent maneuver found in the sociology of knowledge attempts to fully confront the problem of reflexivity. This tactic seeks to use reflexivity as a means for examining the boundaries and limitations of all knowledge, including sociological knowledge (see Woolgar 1982; Woolgar and Ashmore 1988; Ashmore 1989 and 1994). From this perspective, the classic issue of reflexivity is not a problem to be overcome or ignored, but a fresh opportunity to illustrate the limits of all forms of knowledge. It is also seen as an opening to begin experimenting with new, nonrepresentational forms of communication—forms that avoid the pretenses of trying to reflect reality. The reflexivity project's origins can be found not only in Durkheim's more reflexivistic impulse but also in Nietzsche and postmodernism. The project involves immersion within and an experimentation with a variant of strong textual relativism. This sociological version of textual relativism views representation as always problematic and contingent (see Brown 1987; Seidman 1991; Agger 1989).

Here, as in deconstructionism, everything, including society and knowledge and knowledge about society, is transformed into semiotic objects. In turn, these semiotic objects, or texts, are used to show the restrictions imposed by representational forms of discourse. Since the modern epistemic goal of uniting theory and reality is seen as always doomed to failure, sociologists ought to consider the creative aspect of their own writing and should begin to experiment with new literary or nonrepresentational forms of writing. This may include dialogue, one-act plays, self-refuting texts, parallel discourses, parodies, and the use of simulated readers (see Agodi 1991). For example, using multiple voices in a text can be used to remind the reader of the continuous interpretive activity involved in reading a text. As with reader-response theory in literary criticism, multivocality reveals that a text does not have one stable meaning. Rather, meaning constantly shifts and is realigned as a text encounters readers and their interpretative contexts. This means that a text is essentially "unreadable" until it encounters readers and interpreters.

Proponents of the reflexivity project see the examination of reflexivity as being part of a third step in the "natural development of the relativist-constructivist perspective in the social study of science" (Woolgar and Ashmore 1988, 7). The first step, represented most forcefully by Merton, treated both science and social studies of science as realist endeavors. Science discovered natural reality and social scientists discovered social reality. Both were empirical sciences capable of discovering truth within their own separate domains. In the second step, represented by the strong programme and empirical programme of relativism, science becomes relativized, while social studies of science remain realist. In this step, the findings of science are contingent on society, but social studies of science find and report the real, context-free aspects of the sociality of science. In the final step, represented by the reflexivity project and its new literary forms, both science and social studies of science become relativized. Here, both science and social studies of science are recognized to be discursive and constructed. Neither is capable of capturing reality, natural or social. This being the case, it is time to rethink the way theorists go about their intellectual activity.

The reflexivity project's textual relativism can be seen as marking the importation of postmodern theory into the sociology of knowledge. This project, like the postmodernist's, wishes to make all statements, including its own, open to the infinite play of a deconstructionist reading. It is a nonrealist strategy for showing how "human, all too human" all knowledge forms really are. Such a realization forces sociologists and other knowledge producers to admit that they write poetry rather than scientific, representational documents. They must now admit that they are rhetoricians and not realists. In this process, new accounts and methods of knowledge are made possible while sociology and other social sciences are forced to abandon forever their naive social realism.

Conclusion: Accounting for Reflexivity in the Sociology of Knowledge

The sociology of knowledge has historically employed very modern perspectives and methodologies to arrive at what turns out to be, at least in retrospect, very postmodern findings and implications. With the reflexivity project aside, what has united most of the different sociologies of knowledge of the twentieth century is a belief in the explanatory power of some brand of social realism. Sociologists might bicker over the definition of concepts, methods, or theoretical interpretations, but they are in virtual agreement that material conditions, social organization, social interests, social interaction, or actor accounts offer intrinsic meaning and causality. Like their scientific realist counterparts who believe in a natural world beyond consciousness, most sociologists believe in the existence of a social world beyond individual experience and representation. In this view, society always "exists prior to and independently of such [sociological] knowledge" (Layder 1988, 255). The social world, like the natural, can be recovered through the use of methods that, in some way exact, refine, and highlight sensory experience. These methods should be empirical, objective, and rational in order to produce reliable explanations of the social world (see Keat and Urry 1982, 5).

As an antinominalist and anti-idealist discipline, sociology sees itself as going beyond the unanswerable and paradoxical questions of traditional philosophy into something that is pragmatically grounded and empirically or rationally verifiable. Like Durkheim before them, contemporary sociologists see this move away from philosophy and metaphysics as allowing them to make definitive statements on the overall nature of society or some societal issue or problem.

Such an orientation toward the social world certainly has its advantages, at least in an instrumental sense. The taken-for-granted existence of the social, backed by objective methods, allows sociologists to make what they believe to be absolute pronouncements on a range of social issues and problems, including even the content of knowledge. In doing this, however, sociology has slowly opened itself up to epistemological attacks and a general intellectual skepticism about its findings and conclusions.

Although many of the twentieth-century sociological strategies used to explain knowledge have been empirically fruitful, arguably none has been successful in answering or resolving the problems initiated by Durkheim. All these strategies remain, to some degree or another, thoroughly captured by the Durkheimian impulses and dilemmas. Mannheimian and other exceptionalist strategies leave untouched some of the most interesting and powerful forms of knowledge, such as science and mathematics, while remaining unreflective about their own knowledge. Preserving scientific claims from sociological scrutiny does preserve the epistemology of sociology, but it conflicts with the other great impulse of sociology: to examine everything, including epistemology and knowledge, sociologically.

The strategy found in phenomenology and ethnomethodology has proved to be empirically useful, yet it fails to provide any reasons for doing sociology except to retrace and redescribe the movement of actors, however, in the social construction and management of reality. Even in the instance of retracing the movement of actors there appears to be little room for sociological input. If actor accounts are all that constitute reality, why bother to add ethnomethodological ones? Ethnomethodological descriptions are merely another tier of explanation appended to the layer of everyday experience: are they really necessary? Habermas's theory of communicative action also has similar problems. If communication is fragmented into multiple discourses, as postmodernists such as Lyotard claim, how is it possible to have an ideal speech community? Are not competition and conflict also ingredients of communicative action? In addition, on what basis are we to accept the idea that normative impulses are built into communication?

The approaches of the strong programme, social constructionism, and related empirical studies of science and knowledge also seem to have their limitations. The central problem here is the paradoxical retaining of social realism while dispensing with natural realism. As mentioned earlier, advocates of the strong programme and social constructionism grant themselves a realist perspective to describe science and scientists, but they deny such an approach to scientists. They are constructivist where nature is involved and realist where society is concerned (see Latour 1993, 27). Sociologists of science have a "penchant for presenting data to show how irrelevant data actually are (Hull 1988, 4; see Laudan 1982). So, it ironically appears that social realism is more real than natural realism.

The strategy found in the reflexivity project and its strong textual relativism reopens some of the problems found in postmodernism. What is one to do with this continuous exploration of the instability of meaning? After meaning is shown to be unstable by new literary forms, what next? Are intellectual activities supposed to reorganize themselves or does the revelation of the instability of meaning mean that the project of reflexivity is itself unstable and even unbelievable? If the latter position is followed, why not do what was always done methodologically, only maybe this time with a bit more modesty?

In examining the reactions to reflexivity of the different sociologies of knowledge presented in this chapter, I think that is possible to identify three central players and strategies, "specialists," "pragmatic realists," and "reflexivists." The first "specialist," strategy, found in Mannheim and Merton, attempts to make both the natural and social sciences special ways of knowing. Once this is accomplished, there is no reason for epistemological doubt because scientific truth claims are by their very nature better than other types of claims. This approach includes "old-fashioned positivists" who have never lost faith in the goal of developing a truly scientific sociology. Advocates of this view often argue that sociological knowledge can progress if its practitioners learn to emulate more closely the natural sciences, perhaps by formalizing their basic concepts (see Wallace 1995).

The second approach, found in most versions of SSK, ethnomethodology, and social constructionism, often incorporates what might be termed a pragmatic-social realist approach. From this perspective, such theoretical fascination with sociology's own knowledge claims amounts to so much intellectual "navel gazing"; or, in other terms, explaining the act of explaining is an impediment to the practice of explaining (see Blalock 1985). This approach includes sociologists such as Harry Collins and Steve Yearley (1992), who believe that reflecting on the status of sociological truth claims is practically self-defeating. This position heeds the advice of Goffman (1974, 12) that "methodological self-consciousness that is full, immediate, and persistent sets aside all study and analysis except that of the reflexive problem itself, thereby displacing fields of inquiry instead of contributing to them."

Within the pragmatic-realist approach the problem of reflexivity is solved by either ignoring it or by chastising sociology into getting on with the important business of explaining things, such as poverty, racism, and science, while leaving such idealistic quandaries on the status of knowledge to armchair intellectuals, philosophers, and theorists (particularly French) who allegedly have nothing more important to tend to. Pragmatic realism does not seek actually to solve any of the epistemic problems posed by postmodernism or reflexivism; it wants only to bracket them in order to get on with the important work of sociology.

The third response, found in the reflexivity project, takes reflexivism very seriously. It argues that reflexivity is not really a problem at all, but an unparalleled opportunity to reconfigure the entire sociological project—to make it more humanistic and perhaps hermeneutical. In this sense, it is closely allied with the postmodern attempt to dissolve academic boundaries and divisions and to "humanize" all knowledge endeavors. Once this is accomplished, knowledge and truth will finally lose their holy aura.

All of these strategies have been thoroughly criticized both by each other and by approaches outside of sociology. The result has been that sociology still lacks an adequate account of knowledge or a successful strategy for dealing with the issue of reflexivity. Part of the problem here may be that these approaches do little to reconfigure the legacies left by Durkheim or other nineteenth-century social theorists. They all, at least in some manner, harbinger the spell of representation introduced by Durkheim in the late nineteenth and early twentieth centuries. Within these responses, the central dilemma posed by the Durkheimian sociology of knowledge remains inadequately resolved: if knowledge is socially constructed, then must not the sociological knowledge that all knowledge is constructed also be a social construction? If this is the case, "is 'normal science' possible once a constructivist epistemology is taken to heart, or is 'constructivist normal science' an oxymoron" (Fuller 1994, 87); or, as Law (1994, 2) puts it:

> What should we make of this? This is one of the great questions of contemporary social theory, the problem of the status of our writing in the self-reflexive world of high modernity. Do we continue to pretend if only for heuristic purposes that we are

different? Do we therefore insist that social science can tell a reasonable order story of the world? Or do we wrestle with the uncertainties of our own implications and orderings?

Sociologists, like other knowledge producers, "like to think that our theories are better than those of our rivals, that we can see further, that we can find underlying patterns, the deep structures in the social that drive appearances" (Law 1994: 8). The twentieth-century sociology of knowledge is, however, still unable to answer effectively how this is possible. The importation of postmodern theory has only made this task even more complicated and paradoxical.

Notes

1. There are obviously many more strategies that have been employed to answer this question than I consider in this chapter. I am dealing here with the a few of the most common or popular ones. Hermeneutics is another example not discussed here since it is often utilized in several of these overcoming strategies. Hermeneutics, however, suffers from some of the same shortcomings as the ones described. Specifically, it fails to tell us why some interpretations are more successful than others and what makes an interpretation valid.

2. A related approach used in sociology to answer the problem of reflexivity involves the use of metatheory to rid sociological theory of its hidden biases and assumptions (Ritzer 1991). In this approach, social analysis will finally get it right once its philosophical and ideological ghosts have been properly exorcised by a post hoc sociology of sociology. The impetus behind metatheory is, at least in part, the building of better, stronger, more representational theories through the discovery or elaboration of an underlying metalanguage. Allegedly, information regarding the sociological composition of social theories can lead to a more secure foundational language. This language can then be used to reconstruct existing theories, understand more deeply and accurately the theory-building process, or to produce new theory (Ritzer 1991). This approach attempts to solve the problems of relativism and reflexivism by cleansing the discipline of the various tacit ideological assumptions that clutter the royal road to truth.

3. Mannheim later revised this position to some extent in *Essays on the Sociology of Culture* (1956).

4. A similar view of the relationship between science and society can be found in the work of Stephen Cole (1992).

5. For a recent attempt to deal with this issue, see Knorr-Cetina (1994). She (1994, 5) argues, "If science, and modern institutions in general, do not run on facts, this is no rea-

son for despair and resignation—it is rather cause to investigate the ways in which these institutions, if they do not run of facts, run on fictions." It is questionable, however, whether this movement from facts to fictions is successful in solving the reflexivity problem or merely manages to replicate it in a new form.

Chapter Six

Actor-Network and the Construction of Truth

To this point, I have provided basic overviews of three influential current approaches to knowledge: scientific realism, postmodernism, and social realism. Each of these approaches has its own unique ontological understanding of reality and epistemological approaches to knowledge acquisition. Advocates of scientific realism have long argued that reality exists independently of human cognition and representational practices. Epistemologically, this reality can be rationally apprehended if, and only if, the right methods are used to extract and report it. More recently, supporters of postmodern theory have reversed the modern position and its causal arrows between reality and representation by arguing that reality cannot be separated from the signs and symbols used to represent it. In general, postmodernists argue that epistemology, as it was traditionally conceived, is either impossible or at best a constantly changing interpretative activity, much like ascertaining the ever-shifting meaning of a novel. Proponents of the third position, social realism, argue that reality is a culturally variable collective representation of society. Epistemologically, social realists argue that certain empirical methods of inquiry can be used to summon and explain the effects of this social construction of reality.

Throughout the last few centuries these ontological and epistemological accounts have resulted in a rigorous intellectual and political division of labor within academia. In accordance with this division of labor, scientific realists "naturalize" phenomena. They see events as resulting from some inherent natural order of things. In turn, postmodernists "textualize" all phenomena. They see events as a type of storytelling enterprise that results from the linguistic, rhetorical, and discursive activities of human storytellers. Finally, social realists "socialize" all phenomena. They see events as products of social interaction, social hierarchies, social organization, or some other social cause. From the perspective of social realists "the social is the natural" (Lorber 1994, 36).

When placed side by side, these three descriptions of reality and truth constitute the only viable options and possibilities now open to contemporary accounts of knowledge. They also comprise, at least in a simplified form, the three basic positions in the current debate regarding the status of knowledge, reality, and truth. Depending on one's metaphysical (and departmental) position, the cause of knowledge must always be reality, rhetoric, or society. Each of these positions, however, often runs into ambiguity and contradiction. The transcendent naturalism found in scientific realism makes it appear that humans leave few, if any, hermeneutical marks on truth. For many theorists this is difficult to accept because there are numerous historical and contemporary examples of how modern knowledge is influenced by so-called extraepistemic factors. The textualism of postmodernism also leaves many uneasy and dissatisfied. If everything is a text, including the knowledge that everything is a text, then why do some "texts" appear to be more real than others. If all texts are equivalent, then are all truth claims symmetrical? If so, do we abandon the entire truth-finding enterprise outlined in the modern episteme? The all-mighty society of social realism also seems highly problematic and unacceptable. If social forces are really the ultimate cause of knowledge and not a natural order of things or rhetorical storytelling, then that argument also must be a product of society or a group's collective symbolism. If that is the case, sociological accounts can claim no special epistemic status either: they, too, are but a group-specific means of explaining things—of turning chaos into order.

In this chapter, I begin the explication of a fundamentally different approach to reality, truth, and knowledge—that found in the actor-network theory associated with Bruno Latour and Michel Callon. Actor-network theory's origins can be found in some of the sociologies of knowledge considered in the last chapter, particularly ethnomethodology and the sociology of scientific knowledge. Its approach to knowledge, however, consciously seeks to side step many of the problems associated with those sociologies of knowledge. As such, actor-network theory has implications for all the ontological and epistemological views of knowledge, truth, and reality described above. Actor-network theory is critical of scientific realism's privileging of a natural order of things, social realism's emphasis on an absolute social causation, and postmodernism's relativistic and textualistic antirealism. Advocates of actor-network theory seek to develop a model of knowledge that is neither scientific, sociological, or postmodern but is, in the terminology of Bruno Latour (1990), "amodern." Specifically, actor-network theory attempts to negotiate a path between scientific and social realism, which emphasizes the ontological supremacy of nature or society, and an extreme postmodern relativism, which contends that there is no criterion or foundation for determining substantive differences between competing forms of knowledge about reality.

In examining actor-network theory, I focus primarily on the writings of its leading spokesperson, Bruno Latour. His attempts to avoid the ahumanism and apoliticalism of scientific realism, the antirealist textualism of postmodernism, and the reflexivist dilemma of the sociology of knowledge simultaneously operate both substantively and epistemologically. His approach involves both an empirically based network model of how knowledge is created, disseminated, and maintained and a basic reconsideration of the modern foundation or constitution from which he thinks the epistemic problems associated with scientific and social realism and postmodernism originate.[1] As such, it seems to offer a new vocabulary for reconfiguring the spell of representation that has so occupied contemporary accounts of knowledge and has inhibited the development of a postrealist model of truth-building.

In the next three chapters, I look at actor-network theory's unique approach to knowledge, truth, and reality. In this chapter, I focus on Latour's general empirical model of truth making and some of its implications for the understanding and analysis of knowledge. In the next chapter, I explore Latour's views of modernity and modern epistemic dualisms and dilemmas. In the concluding chapter, I summarize some of the implications of his amodernism or nonmodernism for scientific realism, postmodernism, and social realism.

How are Facts Made?

Latour's actor-network theory was initially developed with Steve Woolgar through an in situ account of the daily workings of scientists at the Salk Institute (Latour and Woolgar 1979 and 1986). In the book *Laboratory Life*, Latour and Woolgar use ethnographic methods to illustrate the intricate craftwork used by scientists to construct and fortify truth claims. In the study, Latour and Woolgar deal with scientists in the laboratory much like a field anthropologist would treat members of a traditional society: they attempt to remain neutral about the ongoing attempt of the culture's members to construct and transmit meaning.

Latour and Woolgar argued that science can best be seen as a concerted effort to turn chaos into order. Instead of a scientific object possessing an absolute essence that proper methodology unveils, Latour and Woolgar argue that it is actually the intricate and sustained microsocial work of scientists in the laboratory and their mobilization of resources at their disposal that are responsible for creating and stabilizing the meaning of an object. They contend that a craft object like THF(H), the object of their study, has no meaning except that which is constructed and spread by scientists at work in the laboratory and beyond.[2] Latour and Woolgar (1979, 236) conclude that "reality was the *consequence* of the settlement of a dispute rather than its cause." Reality enters the picture only after the craftwork has been performed.

An elaboration and reformulation of this ethnographical and constructionist approach to science and knowledge can be found in Latour's later works, *Science in Action* (1987) and *The Pasteurization of France* (1988a). Throughout both of these works, Latour seeks to open the "black boxes" of scientific fact production. Like the early sociology of scientific knowledge, Latour argues that this can be done only by following science in action as opposed to looking at ready-made science. In these works, he seeks to examine the numerous processes and associational linkages involved in building and sustaining knowledge. This is accomplished by tracing and following the various links that scientists use in making a network of support and extending a truth claim. Latour contends that in the process of truth building, all statements about the world begin as fragile or weak assertions made by an individual or a small group. At this point, the statement is local, context dependent, contested, and highly rhetorical. In the process of truth making, many statements fall by the wayside while a relative few are elevated to the status of knowledge (see Price 1986). In insolation a "sentence is neither a fact nor a fiction; it is made so by others, later on" (Latour 1987, 25). Assertions are transformed from weak fictions to strong facts as they insinuate themselves into textbooks, introductory lectures, and publications; become part of an organization's collective symbolism; and, perhaps most importantly for science, become inscribed and concealed in experimental hardware and technical apparatuses (see Latour 1987, 103-144; Galision 1987).

It is through this network building that a scientist is able to transform a truth claim from soft, contextual rhetoric to hard, universal truth. Latour argues that what becomes a scientific fact and an accepted truth is a direct outcome of the strength of the ties, or "metrological chains," that a scientist or a group of scientists is able to forge and enforce. If a scientist has been successful in building and maintaining an encompassing network of support, reality has been defined (see Latour 1987, 179). On the other hand, if a scientist has been unsuccessful, the claim fades into the archives of false starts and bad ideas. These processes occur regardless of the intrinsic merits of the claim itself. This means that a statement becomes scientific and truthful only when it stops being isolated and challenged (see Latour 1987, 33). Until the challenges end, the statement is always in danger of being toppled and made false.

In actor-network theory, facts can be constructed only by establishing strong and heterogenous associational networks. For Latour, the stronger and more encompassing the network, the harder a claim becomes. The scientist, by successfully recruiting allies, both human and nonhuman, appealing to authority, referring to former texts and procedures, compiling data, creating computer files, and utilizing laboratories and equipment, is able to create an encompassing network, the sponsorship of a sustaining audience, and, ultimately, a community of truth. The sheer number of human and nonhuman allies, however, is not enough to create, sustain, expand, or defend a network of truth. The truth-making process also requires that these allies be made of different types of things. Heterogeneity, not numbers alone, is what makes a claim potentially strong. For example, in science, readers are asked not just to trust the experimenter, but also to believe their own eyes (see Latour 1987, 66) as well as the gauges and measuring devices of scientific instruments. Likewise, in scientific articles, readers are asked to accept not just the opinions of the author, but also the graphs, charts, statistics, and supporting references of the text. Consequently, truthful scientific networks are made not only of representations or texts, as the rhetoric of science argues, but also of nonhuman components, such as experimental hardware, test objects, and statistics. Nonhuman allies are as important to the success of a truth claim as human ones. Attacking only one component of the network is not enough to defeat a claim.[3] Counterclaims must also be heterogenous so that they are able to fight a knowledge war on all fronts.

In order for a network to be fully established, it is also crucial that the network be able to resist the deconstructionist tactics of its adversaries. In this sense, truth building is seen by Latour as always a politically tenuous activity. While one group is attempting to build and establish a reality order, there are always other knowledge competitors who are trying to do the same. Opponents of a truth claim may use a number of economies of judgment and deconstructionist tactics to attack the credibility of a claim or to convert the claim to their own position (see Fuchs and Ward 1994a). For example, they may focus on such aspects as the reputation of the author making a claim, the procedures and methods used to derive a claim, the theoretical or ideological commitment of the author, or the rhetorical style and coverage of the author. Those claims that do not cover all the bases, so to speak, are more likely to be deconstructed and relegated to the sidelines. For example, if opponents of a claim can illustrate that the promoter of a claim has taken some methodological shortcuts or has an ideological bias toward a particular perspective, the claim can more easily be deconstructed and dismissed. If a claim has insulated itself, however, against such deconstructionist tactics by developing an elaborate and encompassing heterogenous network, it is more likely to find itself transformed into absolute truth—a true reflection of reality.

The final transformation of a statement into a fact occurs when all traces of human agency are erased and replaced with a natural order. In this case, a statement has become so strong that it becomes part of the accepted background knowledge of science and perhaps society at large. In this case, most, if not all, people have been incorporated into the network of truth, including other scientists, high school teachers, media pundits, and the general public. At this point everyone now speaks and is in the truth. Latour reminds us, however, that like having a soul, attention deficit disorder, self-esteem, or karma, being truthful is possible only within the confines of a particular associational network. If an individual steps outside of the network that supports the truthful or is enrolled in another network, he or she believes and, consequently, is something entirely different (see Ward forthcoming).

Latour's Rules of Method

In describing the various associational processes used in the construction of scientific knowledge, Latour also provides seven basic interconnected rules of methodology. He considers these rules as necessities for conducting an actor-network approach to knowledge and truth. In general, these rules can be seen as deliberate attempts to avoid many of the impasses found in the traditional accounts of knowledge, such as those in science, sociology, and postmodern theory.

The first rule of method contends that science should be studied in process rather than ready made. On this point, Latour and the more traditional sociology of scientific knowledge are in complete agreement. Ready made science is what results after a network has been constructed and adversaries subdued. Ready made science is to be seen as cold, hard reality. On the other hand, science in action is much more of an ad hoc and tentative process. Science in the making is warm, soft rhetoric. Science in action does not yet have the allies and network to be called truth. Its success, and therefore its truthfulness, is still very much in doubt. Latour (1987, 14-15) writes:

> We will enter facts and machines while they are in the making; we will carry with us no preconcepts of what constitutes knowledge; we will watch the closure of the black boxes and be careful to distinguish between two contradictory explanations of this closure, one uttered when it is finished, the other while it is being attempted.

This being the case, actor-network theory must accept as facts those rather rare scientific statements that lack controversy, while remaining relativistic with regard to the ongoing struggle of scientists to turn soft statements into hard facts. In Latour's (1987, 100) words:

> If there is no controversy among scientists as to the status of fact, then it is useless to go on talking about interpretation, representation, a biased or distorted world-view, weak and fragile pictures of the world, unfaithful spokesmen. Nature talks straight, facts are facts. Full stop. There is nothing to add and nothing to subtract.

Latour points out that if a position becomes thoroughly relativistic and rejects all knowledge claims, such as those found in postmodernist rhetorical accounts of science, it begins to look absurd since there are issues in the world that have been settled by scientists and other knowledge producers. If we treat science as exclusively rational, however, as in scientific realism or the public talk of scientists, we develop a distorted and otherworldly view of how science produces hard facts and ends controversies. Instead, it is important to develop a "stereophonic" approach to truth making (see Latour 1987, 100). Scientific claims are truthful once they have been stabilized. If, however, there are controversies over the claim, it is still only opinion and rhetoric. This requires that those studying knowledge be Janus faced: both realists and relativists. They are to be realist after facts have been stabilized and thoroughly relativist until stabilization has occurred. In this process, facts are to be seen as nothing more or less than associationally constituted "hard fiction."

The second rule of method states that a truth claim is never the result of the intrinsic qualities of the claim itself, but a consequence of the various transformations the claim undergoes in the hands of others. In other words, a claim can become hard truth only if used and transmitted by other allies. If the claim has no or few allies its truthfulness stops and the claim fades into oblivion. Consequently, advocates of a claim must utilize a number of different truth-building tactics to prosper. One important tactic is rhetoric. Advocates of a claim must successfully incorporate various modes of persuasion to entice or coerce others to their position. Latour writes (1987, 37-38):

> Like a good billiard player, a clever author may calculate shots with three, four, or five rebounds. Whatever the tactics, the general strategy is easy to grasp: do whatever you need to the former literature to render it as helpful as possible to the claims you are making. These rules are simple enough: weaken your enemies, paralyse those you cannot weaken...help your allies if they are attacked, ensure safe communications with those who supply you with instruments, oblige your enemies to fight one another; if you are not sure of winning, be humble and understated.

Echoing in some ways William James's (1991, 88) contention that "truth happens to an idea," Latour argues that "the fate of what we say and make is in later users' hands" (Latour 1987: 29). Consequently, the trick in successful truth construction is not just to enlist large numbers of things and people, but to enroll these allies in such a way that one's fragile statements can become accessible to other allies. This also requires, however, that those allies be strictly controlled so that "what they borrow and spread remains more or less the same" (Latour 1987, 121). Advocates of a claim want their ideas adopted and utilized but only if those ideas are concurrent with their own attempts to spread a claim and forge a truth. To do this, allies must be continuously managed, disciplined, and forced to stay in line. Claims and their supporting allies are always in danger of being retranslated and redeployed to the side of an opponent.

The third rule of method contends that since nature is the outcome of the closing of controversies, it can never be used to explain how controversies are concluded

(Latour 1987, 258). In making this claim, Latour, like the sociology of scientific knowledge, rejects the traditional textbook view of science and scientific methodology found in the philosophy of scientific realism. These realist accounts attribute the success and truthfulness of science to its basic methodology and cognitive style. In other words, nature reveals its secrets to scientists because they have the proper methodological procedures to extract it—their methods are in harmony with nature itself. Latour, however feels that these scientific realist descriptions do not really explain how scientific truths are reached, controversies are settled, or reality is "established and confirmed."[4] They are more the post-hoc constructions and rationalizations of the winners of truth conflicts than accurate descriptions of science in practice. For Latour, the unveiling of nature through scientific experimentation is never the reason a truth claim emerges as victorious. Instead, nature is the final outcome of the settling of controversies. In other words, nature is an effect of the organization of things, not the cause (see Latour and Woolgar 1979, 236; Latour 1987, 98). In Latour's view, nature always arrives too late to "explain the rhetoric of scientific texts and the building of laboratories" (Latour 1987, 94). Nature is the label that results following much knowledge work and after controversies are settled and associational networks stabilized.

The fourth rule of method applies the same principle to the concept of society. It is this argument that clearly differentiates Latour from others in the sociology of scientific knowledge. Since society is also to be viewed as the outcome of the settlement of controversies, it, too, can never be used to explain the success or failure of a truth claim. In this case, the term "society" is merely part of the vocational language of sociology. It can never be successfully used to replace the naturalistic lexicon found in scientific accounts. The "socialization" of objects found in sociology and other social sciences is merely a reversal of the naturalization of objects found in the natural sciences. Latour questions, "How could we take so many precautions in not believing directly what scientists and engineers say about objectivity and subjectivity, and readily believe what other scientists (social this time) say about society, culture and economy" (Latour 1987, 143-144)? Unlike other forms of the sociology of knowledge, which in some way want to privilege the social over the natural, Latour argues that such a sociological approach is always paradox ridden and, in the end, self-defeating.

The fifth rule of method contends that it is imperative for those who study science and knowledge to remain undecided as to what science, knowledge, and technology are actually made of. Since his third and fourth rules of method forbid the invocation of society and nature to explain the success or failure of a truth claim, the focus should now be placed on the strength and type of links that unite objects and people. This requires that we follow the various actants as the network is built and maintained, taking careful note of those who do the work (see Latour 1987, 258). In this move, the focus is placed on network building and not on the entities of society, nature, or the intrinsic qualities of the individual items of the network. This requires that those who study knowledge must simply make a list of the items and links that are part of the network and trace those links across time and space. They should not determine in advance whether these items belong to the domains of nature or society. They should become conceptual atheists and consider objects as nonentities or hybrids, void of essences but still with their own unique and irreducible qualities. Latour (1987, 175) writes,

> From now on, the name of the game will be to leave the boundaries open and to close them only when the people we follow close them. Thus, we have to be as undecided as possible on which elements will be tied together, on which they will start to have a common faith, on which interest will eventually win over which. In other words, we have to be as undecided as the actors we follow.

Thus, instead of looking for natural or social explanations of knowledge, those who study knowledge should remain thoroughly uncommitted about the ultimate meaning and origins of items and links.

The sixth rule of method argues that when irrationality is used to condemn a claim, it is important to look to the network of accusers being built against the claim rather than to some superior reasoning or methodology or the logical structure of a certain social group. In other words, rationality is to be viewed as the cry of the winners of knowledge altercations. Irrationality is not a neutral accusation made in the name of reason against those who have violated the natural order of things or a methodological procedure, but merely a label applied to those who disagree with a particular claim or those who have lost a knowledge skirmish. Rationality and irrationality are, therefore, not built into the cognitive origins of the claims themselves or the claims' ontological origins, but are marks stamped on epistemic losers or transgressions of the moral order of knowledge.

Within the Latourian scheme, the long-standing epistemological question, "Do scientists or moderns have superior access to reality that enables them to discover truth and report knowledge about the world?" becomes moot. For Latour, answering such a question requires one to adopt an asymmetrical view of rational thought and irrational belief, something he and other advocates of actor-network theory are unwilling to do. Traditionally, it has been held that every individual who possesses a "sound mind and a sound method" has the ability to be as rational as the scientist (Latour 1987, 184). Those who lacked this attribute (i.e., the irrational or primitive) were merely under the influence of various social or psychological prejudices. If those prejudices were eliminated, as Francis Bacon and other New Science and Enlightenment writers hoped, the rational potentiality embedded in each person (and indeed all of humanity) would finally flow forth. For Latour, however, it is more productive to think of irrationality and rationality not as states of mind or cognitive typologies, but as associationally constructed insignias. Irrationality, like rationality, has to be constructed through networks—in this case a strong network of accusers. Rationality and irrationality are merely the resulting designation "of an attack by someone on associations that stand in the way" (Latour 1987, 205).

For example, if irrationality is "put on trial," the results are always ambiguous. For every claim of irrationality in one knowledge system, there exists an element of irrationality in the "rational" system doing the accusing. It turns out that "everyone on earth is as logical or as illogical as any one else" (Latour 1987, 195). Latour (1987, 196) concludes that "no asymmetry between people's reasoning can be recognized." Essentially, everyone is a practical or mundane reasoner employing various "ethnomethods," to use a term from ethnomethodology, to construct reality (see Garfinkel 1967). Consequently, every reasoner, ordinary or professional, modern or nonmodern, scientific or humanistic, is as rational or irrational as the next person. In scientific reasoning, as in a court proceeding, those who win a case become the truthful and the just.

The seventh and final rule of method is an outgrowth of the sixth one. It asserts that searches for the cognitive characteristics that make some claims more special than others be suspended. Since the development of the modern episteme, philosophers have argued that truth can be differentiated from myth by the cognitive characteristic of those who produce each type of claim. Yet, for Latour those attempts that resort to cognitive explanations to explain the special characteristics of certain types of knowledge are never successful. On closer examination knowledge has nothing to do with the cognitive attributes of those who produce it. This is so because rationality is not inherent to certain minds, but a term applied after the fact to successful claims.

If reasoning, however, is not a matter of qualitative cognitive differences between people, as traditional realist epistemology argues, why do we not all share the same beliefs? By eliminating the cognitive distinction between rationality and irrationality, Latour is able to provide an answer to this question. Like the scientist in the laboratory, the practical or mundane reasoner of everyday life is concerned with "what can be tied to a claim to make it stronger and how can the claims that contradict it be untied" (Latour 1987, 198). Consequently, building reality is both a constructive and deconstructive process. Essentially, people think neither logically nor illogically—they think sociologically, or associationally. All people move from element to element as they construct statements about the world. When a controversy starts, "they look for stronger and more resistant allies..." (1987, 205). Some are able to recruit more allies than others and, as a consequence, their truths become harder.

What separates science from everyday practical reason or belief is not simply an issue of logic or types of cognition, but an issue of the level of controversy and the appropriate associational ties needed to be mobilized in order to end the controversy. Essentially, the difference is a matter of content rather than form. The everyday practical reasoner does not have to mobilize as many allies in the production or replication of standard taken-for-granted knowledge as the scientist does for the production and defense of a hard fact. As evidence, Latour (1987, 206) provides the following example:

> 'An apple a day keeps the doctor away,' the mother said handing out a glowing red apple to her son, expecting a grin. 'Mother,' replied the child indignantly, 'three NIH studies have shown that on a sample of 458 Americans of all ages there was no statistically significant decrease in the number of house calls by family doctors; no, I will not eat this apple.'

In mobilizing the statistics of the National Institutes of Health, the child is acting as if this everyday colloquialism is in need of the same type of support employed by science. In other words, the child is treating this soft fact as if it were a hard one and in need of the same type of documentation and class of supporting allies. Since this colloquialism has passed from generation to generation without being challenged, it becomes true. There is no need to treat it as an irrational belief or as a potential scientific fact. Essentially, soft facts are basically all we need for the conduct of everyday life. Everyday life does not require the same types of mobilizations or allies as the hard knowledge of science.

Latour argues that when in the course of human events we answer such questions as: "How much money did I earn this month?" "Is my blood pressure above or below normal?" "Where was my grandfather born?" "Where is the tip of Sakhalin Island?" we can provide either soft or hard answers, depending on who is doing the questioning (Latour 1987, 252). If we provide hard answers, we must appeal to supporting documents and begin the process of stacking up hard facts and constituting hard truth. Latour (1987, 252) writes:

> Even the question 'who are you' cannot be solved, in some extreme situations, without superimposing passports to fingerprints to birth certificates to photographs, that is without constituting a file that brings together many different paper forms of various origins. You might very well know who you are and be satisfied with a very soft answer to this absurd query, but the policeman, who raises the question from the point of view of a centre, wants to have a harder answer than that...

In short, soft answers usually do not require the same type of metrological chains as hard answers. They differ not by their degree of logic, but by the type of networks required for their construction. They can be seen as tribe-specific measures and practices for the accumulation of soft truth.

Why Science is Stronger than NonScience

Utilizing the ideas from Latour's discussion of fact making and his rules of methodology, particularly his discussion of the differences between rationality and irrationality, it is possible to provide an explanation of the divergence and hierarchicalization of scientific and nonscientific forms of knowledge (e.g., literature and lay knowledge). As discussed in chapter 1, within the modern episteme, scientific knowledge is separated and given supremacy over literary and lay knowledge. One of the central reasons for this elevation of science over nonscience is said to be the unparalleled power of scientific methodology to reveal reality compared with the mere speculative force of nonscientific discourse. For actor-network theory, however, it is not the quality of claims that separate literary and other nonscientific forms of knowledge from science, or is the separation the result of an ontologically based division of labor between things of nature and things of culture. Instead, these differences are a result of the types of allies enlisted and the power of the network the allies are able to form. Simply put, science is an associationally stronger, not better, form of knowledge than literature, poetry, lay knowledge, or art.

For Latour, one of the most important factors that distinguishes scientific from nonscientific knowledge is the type of allies that are recruited into a network of support. Scientists recruit not only people and their textual wares, but also recruit and incorporate nonhumans.[5] Nonhumans, such as DNA, microbes, and machines, become part of the network that supports and sustains scientific knowledge. In this sense, nonhumans are seen not as passive objects waiting for scientific manipulation or representation, but as active and contributing members of an associational network (see Callon 1987). As objects become recruited into a network, they make the network more solid and truthful. Their existence within a network makes the network stronger and more resilient.

Another important factor that distinguishes science from other forms of knowledge is the degree of material and human resources available for different groups in the construction of facts and knowledge. Latour (1987, 179) writes:

> The proof race is so expensive that only a few people, nations, institutions or professions are able to sustain it...this means that the production of facts and artifacts will not occur everywhere and for free, but will occur only at restricted places at particular times.

Essentially, strong forms of fact production are relegated to those times and places that can garner the resources and networks capable of transforming weak statements into strong facts. Like making steel or an automobile, this means that production must be centralized at specific sites of production or centers of calculation, such as at think-tanks, corporate headquarters, research institutes, or bureaucratic agencies (see Ward 1995a). It is only at these centers of calculation where people and things can be properly mobilized and controlled to produce knowledge. This does not mean that knowledge that cannot muster the resources is methodologically invalid or epistemologically irrelevant, it is simply not officially sanctioned or hard knowledge.

In general terms, scientists hold an advantage over literary and other humanist intellectuals in building knowledge. Scientists are able to rely on a host of nonhuman allies, such as graphs, statistics, experimental data and hardware, large research centers, large research funds, and the support of governmental agencies such as the NSF and the NIH in the promotion of their truth claims. On the other hand, literary intellectuals usually have only their fragile statements and the citations of the fragile statements of other literary intellectuals. Consequently, their claims are not very heterogeneous or encompassing. They easily break down when one element of the network is challenged. The lack of diversity limits the scope of their knowledge claims. Since science has defined the rules by which legitimate knowledge is to be evaluated, the arguments made by literary intellectuals can be more easily relegated to the sidelines. Literary or nonscientific intellectuals are simply not able to assemble the flexible and all-consuming coalitions needed to forge truth, at least the hard version of hard truth accepted as legitimate in science. A weak network, such as the one constructed by literary intellectuals, has historically had little chance of becoming accepted as fact since it is limited in range and does not adhere to the rules established by scientists as to what will count as truth, verification, and knowledge. It bears repeating that for actor-network theory this has less to with do the type of claims made by the literary academic or the intrinsic merits of the arguments themselves than with the associationally enhanced power of the individuals and organizations making the pronouncement.

Such an argument is also relevant to deconstructionist and rhetorical accounts of science provided by postmodernists. By treating everything as a text or semiotic object, deconstructionists arguably miss other key components of the network of science (see Latour 1987, 63). Consequently, Latour concludes that semiotics is "inadequate because it persists in considering only texts or symbols instead of also dealing with things in themselves" (Latour 1988a, 183). It, like all literary knowledge in general, is limited by its recognition of only texts as sources of meaning.

Conclusion: An Amodern Model of Fact Making

John Law (1994, 100) has described actor-network theory's empirical approach to knowledge as symbolic interactionism "with an added dash of Machiavellian political theory, a portion of (suitably diluted) discourse analysis, and a commitment to the project of understanding the *material* character of the networks of the social." Such a description, while helpful in situating the origins of actor-network theory, makes the position appear as if it were just another version of the more standard sociology of knowledge examined in the last chapter. As such, the description perhaps does not do complete justice to the distinctiveness or radicalness of the approach. While actor-network theory is undoubtedly indebted to the perspectives described by Law, it does seem to offer at least three inventive insights into knowledge that go beyond previous sociologies of knowledge and are, therefore, worthy of further consideration.

Actor-network theory's first contribution can be found in its argument that truth should be treated associationally and not logically or cognitively. Here, Latour introduces a move from a dualistic logic that creates binary divisions between us and them (e.g., the rational and the irrational, the modern and the primitive, the scientific and the literary, and the professional and the lay) to a multifaceted collective logic that looks to the specific network of knowledge practitioners. The logicoempirical model of traditional epistemology has been concerned with drawing and sustaining distinctions between rationality and irrationality, text and reality, modern and primitive, human and nonhuman, and knowledge and belief. These divisions are thought by epistemologists to derive from distinct ontological states. Philosophers only label and report these naturally derived states of reality, they do not create them. Latour and his actor-network theory make no such modernistic (or postmodernistic) distinctions. All distinctions are equally logical or illogical. The inherent quality of various forms of discourse cannot be clearly distinguished. While they are not philosophically or epistemologically distinguishable, they can be differentiated by the types of strong or weak associations evident. In other words, belief and knowledge are separated by the degree of associational practices involved, not by their source or referent.

Second, actor-network theory requires that those who want to understand knowledge-making drop the externalistic and reductionist view of knowledge found not only in the naturalistic accounts of scientists, but also in the "socialistic" ones of social scientists. While such a position is also found in other positions in sociology, such as phenomenology and ethnomethodology, it is found there only in a very limited form. In fact, as I argued in the last chapter, these two positions still retain a privileged view of the social, this time in the guise of subterranean privileged actor accounts. As a result, they also suffer from the classic problem of reflexivity. In response to the Durkheimian dualism between knowledge and society, Latour argues that society cannot be the cause of knowledge since both knowledge and society are the coproductions of actant activity (see Callon 1987; Latour 1991). Knowledge work is always social work. In Latour's view there are no entities of society or nature to cause things. Nature and society are merely the results of actant activity.

Introducing the role of the network in truth building allows Latour to develop his third unique contribution—a nonobjectivistic and a nonrelativistic (or a nonmodern and a nonpostmodern) vision of truth. In both the modern and postmodern cases, truth is not allowed to vary. Truth is either on or off. We either have it or we don't. For modernists, we have absolute truth if we do things correctly. For postmodernists, absolute truth is always off. Truth is impossible since we cannot escape the prison house of language, rhetoric, and discourse. In Latour's account, even that "august thing" called truth is allowed to vary. At some points we have truth, at other points we have partial truths (i.e., truth in the making or truth being deconstructed), and at others times we do not have truth at all. Truths are to be analyzed by their stages of becoming or destruction, not by their intrinsic nature or correspondence. For Latour, if truth claims are not attacked and there is no controversy among competing truth providers, then the claims are true. In his (1987, 206) words:

> If they are not attacked, people know exactly what nature is; they are objective; they tell the truth; they do not live in a society or a culture that could influence their grasp of things, they simply grasp things in themselves; their spokespersons are not 'interpreting' phenomena, nature talks through them directly.

If the so-called black boxes of knowledge are sealed, people do not "live in a world of fiction, representation, symbol, approximation, convention: they are simply right" (1987, 206). If there is a controversy, various strategies are employed to strengthen the rhetoric into a hard fact and to forge another black box. From this position, what distinguishes irrationality from rationality, belief from science, or text from reality is not the cognitive level or type of the participating actants, but the associational enhanced power of some individuals and groups to establish resistant coalitions. Those who do are elevated to the labels of the rational, the scientific, and the discoverer of reality. Those who fail are the irrational, the believers, the followers.

Latour (1987, 179) uses these insights to redefine reality as "what resists all efforts at modification." From this perspective, there is no difference between the real and the unreal, only differences between those coalitions "that resist for long and those that do not, those that resist courageously and those that do not, those that know how to ally or isolate themselves and those that do not" (Latour 1988a, 159). In this connection, "nothing is known—only realized" (Latour 1988a, 159).

In summation, the networks of heterogeneous actants can be conceptualized as serving as "the railroad industry of epistemology" (Haraway 1991, 248). Furthermore, facts can only "run on the tracks laid down" by the network (Haraway 1991, 248). Statements become truer as the associational network and its constructions expand, recruit more actants, enroll more allies, and become more encompassing. In this sense, truth has more in common with a cable television network than with the accounts provided in traditional epistemology. One must plug in other actants in order to construct a network of the truthful (Latour 1993, 119). Conversely, one must be plugged in in order to be in the network of the truthful. What is more, truth is always limited to, and can never extend beyond, the confines of the network that actants have forged. A truth claim's impact and truthfulness are forever dependent upon the length of the connection and the size of the network constructed. This allows Latour to make the seemingly contradictory argument that scientific facts can be both constructed and true.

Notes

1. The importance of the actor-network theory for the discussion here does not necessarily come from the specifics of its discussion of science and technology, but from the general approach to knowledge it seeks to develop and the strategy used to explicate it. Thus, I do not consider the extensive case studies that are used in Latour's discussions.

2. It is significant that Latour and Woolgar lose the word "social" from the subtitle between the first (1979) and second (1986) editions of *Laboratory Life*. This is undoubtedly representative of both Latour's and Woolgar's rethinking of the primacy of sociological accounts of knowledge.

3. Latour uses this notion of heterogeneity to explain how Western nations were able to colonize other cultures. Essentially, it was their heterogeneity that made them strong. Simply attacking the church and its missionaries or the army and its soldiers would not defeat the Westerners since their systems were related but nevertheless distinct.

4. The phrase "constructed and confirmed" is from Bourdieu, Chamboredon, and Passeron (1991).

5. The inclusion of nonhuman agency in network building is one of the key distinctions between actor-network theory and SSK and social constructionism. It is also one of actor-network theory's more controversial ideas. For a critique of the inclusion of nonhumans in accounts of scientific knowledge, see Collins and Yearley 1992.

Chapter 7

Actor-Network's Amodernism

If we were to stop with only an examination of Latour's empirical model of truth making, we would not have a very complete description of his overall theoretical project or how the project is related to the current modernism/postmodernism debate. With only the previous description as a guide, it could easily be argued that Latour's actor-network approach to knowledge is merely a politically astute and sociologically thick consensus theory of truth. It could also be maintained that its central concept of actor-networks, like the concepts found in other sociologies of knowledge, maintains a form of sociological reductionism and is, therefore, also haunted by the classic problem of reflexivity. If, however, one examines Latour's later writings, particularly his more theoretical work *We Have Never Been Modern* (1993), these charges appear to be inaccurate or, at least, incomplete. Latour's goal is not only to provide an adequate empirical model for explaining knowledge, but also to provide a new strategy to bypass the epistemic dualisms and paradoxes that he believes have haunted modern thought and the epistemological projects of scientific realism, social realism, and postmodernism. In fact, it can be argued that an adequate empirical model of knowledge is possible only after the epistemic and political divisions of modernity have been reexamined and partially dismantled. If this is not done, all accounts of knowledge, whether they be empirical, rational, realist, social, or postmodern in orientation, will remain paradoxical and impotent. In this case, the knowledge wars between modern realists and postmodern relativists have the potential of continuing ad infinitum. As we shall see, Latour's writings provide a number of central insights that have an important bearing on the current debate over knowledge, truth, and reality.

In the previous chapter, I examined actor-network theory's basic approach to truth making. In this chapter, I consider Latour's more general analysis of modernity and modern and postmodern ways of knowing. I want to illustrate how such an approach led Latour and other advocates of actor-network theory to call for a complete reformulation of modern and postmodern ontological positions and how these reformulations led to the development of something he calls the amodern or nonmodern constitution.[1] In order to do this, I first examine Latour's reading of what he labels "the modern constitution." Next, I consider how the proliferation of what he refers to as nature/culture hybrids have made the modern constitution increasingly less tenable. Third, I examine Latour's proposals for a nonmodern constitution, as well as how this constitution seeks to keep the positive attributes of the modern way of knowing while ridding it of some of its more problematic and clandestine elements. Finally, I consider some of the general implications of the modern constitution for contemporary knowledge practices and how it can be seen as being responsible for framing the either/or options of both realism and relativism.

The Modern Constitution of Truth

Scientific realism, social realism, and postmodern textualism all share the conviction that modernity marks an unprecedented shift in the Occidental way of knowing. All agree that there is something unique and atypical about the way modernity organizes and conceptualizes reality and produces truth. In doing so, all of these positions, at least in some manner, posit a great epistemic divide or break between the modern and premodern or Western and non-Western ways of knowing. For example, both modernists and postmodernists treat modernity as a reflection of an ultimate sociocultural break rather than as a constructed concept. From this perspective, modern society marks an unparalleled divergence from traditional forms of society and previous ways of knowledge. Latour's discussion of modernity, however, seeks to avoid thinking about modernity in terms of great divides, epistemic breaks, or reflections of a remarkable changing social reality. He argues that the modern approach to reality and truth did not emerge simply as the result of some large Foucauldian-described epistemic upheaval or Kuhnian-depicted paradigm shift. From his perspective, there are no Copernican revolutions to start it all or Hobbesian leviathans to overcome. Instead, Latour (1993, 76) argues that "it is the sorting that makes the times, not the times that make the sorting." In other words, the basic empirical model developed in actor-network theory leads Latour to reformulate the inherent use of structuralistic explanations to ground knowledge, such as those found in other modern approaches to knowledge. The question for Latour is not how some linguistic, social, or natural structure affects the ordering of things, as in postmodernism, social realism, and scientific realism respectively, but how the sorting is achieved, how networks of people and things cohere and stabilize, and how objects come to be assigned, sorted, choreographed, purified, and relegated to the poles of nature or society. As with his rules of method, Latour strives to remain neutral about whether the items that constitute the network of modern knowledge are, by their intrinsic qualities, of social, natural, or discursive origin. Consequently, modernity can be defined by its unique system of sorting and purifying objects, not by its new revolutionary approach to knowledge.

Latour's explication of an amodern ontology begins with a consideration of what he refers to as "The Modern Constitution" (1990; 1993). The modern constitution can be thought of as the grand epistemic law book of modernity, complete with prescriptions and regulations for sorting and purifying objects into distinct ontological and epistemological domains. Like a political constitution, this constitution of truth "allocates the rights, duties, appeals, and branches" (Latour 1990, 156) of all entities, objects, and subjects under its domain. It defines "what Nature is supposed to be, what women are allowed to feel and think, the way labourers are allowed to behave, how God is supposed to intervene and rule" (Latour 1990, 156).

Latour argues that this modern constitution of truth, like all political constitutions, provides certain basic guarantees. He identifies four basic guarantees that serve as types of governmental rules and as mechanisms of checks and balances for the modern constitution. These guarantees only "had meaning when they were taken together but also kept strictly separate" (Latour 1993, 139). The first guarantee made nature transcendent but at the same time allowed for its mobilization. The modern constitution of truth argued that nature was an entity that existed beyond human activity. In other words, nature held a reality that was independent of what people thought about it. Since, however, a completely transcendental nature would make human comprehension and intervention impossible (i.e., knowable and manipulatable only by a transcendent god), it also maintained in practice that nature could be both captured by methodological intervention and harnessed for practical use. The second guarantee of the modern constitution maintained that society was open to human manipulation and alteration. Society was, therefore, something that humans continuously made and remade. Since such an absolute constructionist view of society would make social order problematic, however, it also maintained that society was transcendent and all-determining. The third guarantee of the constitution was that nature and society were two completely separate and distinct domains with their own unique ontological objects. This third guarantee ensured that the first two guarantees would be less contradictory by dividing objects and, as a result, knowledge into two distinct spheres. The fourth guarantee removes God from a direct influence on society and nature but also leaves this "crossed-out God" as the ultimate arbitrator of the assignment of objects to the poles of society or nature.

Consequently, the modern constitution can be seen as a paradox-ridden document filled with a number of constitutional contradictions. These contradictions often prove to be beneficial, however. One of the chief strengths of the modern constitution, in Latour's view, is that it provides a number of articles that intermingle notions of transcendence and immanence. Nature and society are to be seen as simultaneously both transcendent and immanent entities. Nature is beyond human affairs, but also capable of being harnessed and manipulated by humans. Likewise, society is within our power to make, but also takes on the characteristic of something beyond and superior to us. Finally, God is "protestantized" by being removed from both daily affairs and the workings of nature and society. Yet, this God exiled to heaven is also capable of being used as an arbitrator of the ultimate nature of objects. The power of such an arrangement resides in its ability to make all things possible. Latour (1993, 34) describes these intermingling tenets of the modern constitution as:

> A threefold transcendence and a threefold immanence in a crisscrossed schema that locks in all possibilities. They have not made Nature; they have made Society; they make Nature; they have not made Society; they have not made either, God has made everything; God has made nothing, they have made everything.

It is this mixing of transcendence and immanence and the resulting ability to compartmentalize and explain everything that gives the moderns their power over premoderns or non-Western orderings. In Latour's account, premoderns mix things and people together into one general "nature-culture." They often do not have divisions between people and things, political power and natural phenomena, or transcendence and immanence. Nature and society are always interconnected and mutually dependent. They are unable to change the "social order without modifying the natural order—and vice versa" (Latour 1993, 42). The moderns, on the other hand, do have a vast and heterogenous system of ontological government. They are capable of both dividing and explaining. In fact, much of their power derives from this separation of realms and the resulting flexibility that this division produces. The moderns are capable of changing one domain without affecting the other.

In fact, the constitution the moderns develop is so flexible that all questions can be answered within its framework and by employing its basic guarantees. It is, therefore, always able to subdue and redirect its critics. Latour writes (1993, 37),

> If you criticize them by saying that Nature is a world constructed by human hands, they will show you that it is transcendent, that science is a mere intermediary allowing access to Nature, and that they keep their hands off. If you tell them that we are free and that our destiny is in our own hands, they will tell you that Society is transcendent and its laws infinitely surpass us. If you object that they are being duplicitous, they will show you that they never confuse the Laws of Nature with imprescriptible human freedom...

The modern constitution of truth, then, can be thought of, at least at one level, as a well-developed, heterogenous, and comprehensive associational network. It simultaneously allows and disallows. It takes care of contradiction by creating a firm intellectual and academic division of labor and by establishing filtering mechanisms to sort and purify objects. When an opponent challenges one of its positions, defenders of the constitution can easily take refuge in another one of its basic guarantees. For example, natural scientists have both the transcendent state of nature and the power of the human mind to draw upon if needed. Likewise, social scientists have both the society as god of structuralism and the freedom of social constructionism and phenomenology to summon if needed. Each group has the ability to draw on a transcendent or immanent entity depending on what the situation calls for. In the end, it is the flexible and paradoxical characteristics of the modern constitution that have made it so encompassing and successful.

Within this constitution of truth and its guarantees, the most important divisions or branches of government are nature and society. It is these divisions that allow the constitution to work effectively without being overcome by the contradictions found in the first two guarantees. In the modern constitution, there are, on the one hand, those things/objects that are said to belong exclusively to the realm of nature, such as microbes, DNA, black holes, or anything nonhuman.[2] On the other hand, there are those things/objects that are said to belong exclusively to the domain of society, such as politics, interests, ideologies, citizens, or anything human. Within this modern constitution there also exist special methods and procedures that are said to capture nature, while there are other unique methods thought capable of capturing society and culture. When disputes arise over the proper domain for the assignment of objects, the constitution resorts to its purification and mediation processes, somewhat like epistemic courts of appeal. The purification and mediation processes make sure that each object is carefully relegated to its proper nature or society pole. These processes create a division between things that humans cause or influence and those they do not. The object of investigation can only be of nature or of society—never both.

Latour uses the cases of Boyle and Hobbes, borrowed from a critical reading of Shapin and Schaffer's *Leviathan and the Air-Pump* (1985), to illustrate the seventeenth-century struggle to construct this modern constitution of truth and its great divide. On the one hand, Hobbes's "political science" sought to construct a political culture beyond nature. Here, nature and nonhuman activity can never have a bearing on what people do in the pure realm of political culture. Culture was seen as a purely human realm influenceable only by the creative practical and political activities of human actors. On the other hand, Boyle's "scientific politics" endeavored to construct the rules and rituals for the discovery of nature through scientific experiments. Nature and nonhumans and their rules were beyond the reaches of political culture and human cognitive and practical alteration. Nature was a purely nonhuman realm influenceable only by an absolute nature of things. Boyle's experimental science also introduced the intervention of inert bodies or nonhuman actors into knowledge building. These bodies become active components of knowledge making: They are "capable of showing, signing, writing, and scribbling on laboratory instruments before trustworthy witnesses" (Latour 1993, 23).

Comparing Hobbes's and Boyle's efforts to construct the modern constitution, Latour (1993, 28) writes:

> They are like a pair of Founding Fathers, acting in concert to promote one and the same innovation in political theory: The representation of nonhumans belongs to science, but science is not allowed to appeal to politics; the representation of citizens belongs to politics, but politics is not allowed to have relation to the nonhumans produced and mobilized by science and technology.

From Latour's perspective, both Hobbes and Boyle are essentially "inventing our modern world" (Latour 1993, 27) and its epistemic and political divisions between society and nature, geisteswissenschaften and naturwissenschaften, and humans and nonhumans. They have provided the resources of "social force and power" on the social divide and "natural force and mechanism" on the other (Latour 1993, 29). Hobbes's social vocabulary, and later the lexicon of social scientists, becomes filled with culturally descriptive words such as representation, sovereign, contract, property, and citizens, while Boyle's naturalistic lexicon is permeated with scientific terms such as experiment, fact, evidence, observation, and peers (see Latour 1993, 25). For Latour, it is only after Boyle and Hobbes have divided the world and settled their epistemic differences that the modern world begins to take shape.

Hybridization, Quasi-Objects, and the End of the Modern Constitution of Truth

For Latour there is reason to believe that the modern constitution of truth that makes possible the various ontological possibilities and epistemic options of modernity is becoming less tenable or perhaps even never truly existed. In his view, the modern constitution of truth is composed of two sets of divergent but interrelated and mutually dependent practices—translation and purification. The first practice of translation "creates mixtures between entirely new types of beings, hybrids of nature and culture" (Latour 1993, 10). The second practice establishes two discrete "ontological zones" between humans and nonhumans (Latour 1993, 10). The guarantees of modernization required the assignment of objects to the binary opposition of either nature or society. At the same time, however, other guarantees created hybrids that could not be explained by resorting exclusively to naturalistic or socialistic explanations. In fact, Latour argues that the two founding fathers of the modern constitution, Hobbes and Boyle, go right on mixing things (e.g., humans and nonhumans, texts and objects, and nature and culture) and creating hybrids even as they are developing the distinct ontological boundaries and territories of the modern constitution. So, although the third guarantee of the modern constitution tried to separate items cleanly into social and natural domains, the first two guarantees ensured that hybrids would always proliferate. As much as the third guarantee tried to control the creation and growth of hybrids, its success was always limited by the contradictions contained within the constitution itself.

Latour argues that in our own time, we are surrounded by strange hybrids or quasi-objects combining technology and society, human and nonhuman, politics and science, and nature and culture in new and indistinguishable ways. In these hybrids, science is never completely separated from culture. All that is visible are stronger or weaker associations and linkages (see Shapin 1988, 539). Issues such as ozone depletion, the AIDS epidemic, DNA mapping, the abortion debate, and artificial intelligence can no longer be easily purified, mediated, and relegated to the poles of society or nature. For example, "the smallest AIDS virus takes you from sex to the unconscious, then to Africa, tissue cultures, DNA and San Francisco" (Latour 1993, 2). The truth of these issues and objects lies in their mixed, nonessential origins, not in their distinct ontological status. The networks that make objects such as AIDS and DNA real are so interwoven with objects, things, people, and texts that it is impossible to separate neatly the ingredients of the network into aspects of nature or culture. It is no longer possible to argue that AIDS or DNA belongs exclusively to the realm of nature or culture: AIDS and DNA are completely naturalized and socialized. Latour (1993, 6) writes,

> The ozone hole is too social and too narrated to be truly natural; the strategy of industrial firms and heads of state is too full of chemical reactions to be reduced to power and interest; the discourse of the ecosphere is too real and too social to boil down to meaning effects.

For Latour (1993, 7), things are always "simultaneously real, social and narrated." It is no longer possible to look only at nature, political/social interests, or rhetoric and discourse to explain knowledge. In other words, the mediational and translational practices of modernity have broken down. Just as today's courts are unable to process all of their cases, the modern constitution of truth and its divisions and poles are no longer capable of explaining or containing the complex hybrids found in contemporary life. As a result, the modern constitution becomes simply no longer effective.

Yet, the demise of the modern constitution of truth is not due exclusively to the current visibility of nature-culture hybrids. Latour also believes that this mixing of people and things has always been the means through which networks have been mobilized and stabilized. Hybridization is, therefore, not only a characteristic of contemporary life, but part of an ancient anthropological means of organizing. Latour (1993, 107) writes:

> No one has ever heard of a collective that did not mobilize heaven and earth in its composition, along with bodies and souls, property and law, god and ancestors, power and beliefs, beasts and fictional beings....Such is the ancient anthropological matrix, the one we have never abandoned.

This leads Latour to argue that "we have never been modern." Mixing and hybridization have always been a part of creating and defining the order of things. Knowledge has always fused an assortment of elements in order to forge truths and explanations. It is only modern knowledge that is obsessed with the mission of sorting all elements into absolute ontological zones and causes.

For Latour, it is only now that hybridization and the sorting process are becoming manifest and our amodernism begins to show. It turns out that we have always been amodernists. We have always mixed things and people together in order to build associational networks that support knowledge, reality, and truth. What separates moderns from premoderns is that the modernists could not tolerate ontological disorder or matter out of place. They were obsessed with establishing ontological zones and with assigning objects to their proper place. It turns out, however, that the modern constitution of truth, despite its explanatory power and persuasiveness, was never able to control and discipline completely the unruly mob of quasi-objects it created.

The Nonmodern Constitution of Truth

For Latour, a different configuration or representation of knowledge, truth, and reality is needed that does not adhere to the separation of nature and culture, humans and nonhumans, and science and society. In his view, it is imperative that a new account of knowledge be developed that avoids the modern distinction between "things in themselves and humans among themselves" (Latour 1991, 9). None of the great schemes of modernity, such as naturalism, sociology, phenomenology, structuralism, poststructuralism, or deconstructionism, has proven able to overcome the contradictions and paradoxes of the modern constitution of truth. This is so because they are essentially part of its divisions and boundaries. They have been diligent players in the division of the world and its objects into the natural and the cultural. In this sense, the current debate between "naturalists," "socialists," and "textualists" has not solved any of the problems associated with the modern approaches to knowledge. These positions have provided answers only from within their own ontologically separate realms of knowledge. The ensuing debate between these camps over what is really responsible for reality has led only to an epistemic constitutional crisis rather than a solution. Latour (1993, 128) writes:

> No, we do not fall from Nature into the Social, from the Social into Discourse, from Discourse into God, from God into Being. Those agencies had a constitutional role to play only so long as they remained distinct. No one of them can cover, fill, subsume the others; no one of them can serve to describe the work of mediation and translation.

With the proliferation of hybrids and the demise of the modern constitution of truth and its classifications and sorting procedures, it arguably becomes possible to provide an account of truth and reality that is not trapped by the contradictions of modernism or postmodernism—of searching for natural or social facts or for culturally constructed fictions.

In order to provide such an account, Latour wants to reestablish symmetry between the two branches of the modern constitution of truth, society, and nature (see Latour 1993, 138). To do this, he advocates the creation of a nonmodern constitution. This nonmodern constitution strives to maintain the positive features of the modern constitution while discarding its paradoxical, self-defeating, and clandestine components. The nonmodern constitution, like its modern counterpart, also has four basic guarantees (see Latour 1993, 141). The first guarantee is that of the coproduction, irreducibility, and nonseparablity of society and nature (Latour 1993, 141). Under the nonmodern constitution, all objects are to be treated as quasi-objects. These quasi-objects are not to be considered as the results of society or nature but as causes of them. As with his empirical model of fact making, it is the stabilization of these quasi-objects that results in the things called nature and society. In fact, these objects only have meaning once they have jelled into their social or natural form. As with all constitutions, this nonmodern one also has certain moral prescripts. In the case of the first guarantee, the moral prescripts demand that "every concept, every institution, every practice that interferes with the continuous deployment of collectivities and their experimentation with hybrids will be deemed dangerous, harmful, and—we may as well say it—immoral" (Latour 1993, 139). In other words, the nonmodern constitution should prohibit any realm for offering absolutist and reductionist claims.

Latour does not, however, advocate an approach to knowledge that in some manner returns to premodern, holistic, nature-cultures. Nature-cultures were always restricted by their absolute interconnectedness of things. They do not allow changes to be made in one realm without changes being made in another. They therefore have the consequential limitation of restricting "experimentation on a large scale" (Latour 1993, 140). For example, because there is no distinction between nature and culture, if the group desires to intervene in nature it must always alter its society. To avoid this time-consuming restriction on investigation of the premoderns, he calls for a second guarantee. This guarantee advocates the production of an objective nature and a free society. Here, the treatment of nature as transcendent and society as immanent turns out to be a virtue of the moderns worth salvaging, at least after the first revision is implemented. For Latour, it is desirable that nature be transcendent and society be immanent. There should be scientific facts and a society capable of altering and reconstructing itself. Transcendent natural facts allow for truth, while an immanent society permits cultural change and amelioration. All that has happened in the nonmodern constitution is that the coproduction of society and nature have been brought to the forefront and into full view. As scientists use quasi-objects to construct nature, they make it transcendent. In this sense, the malleable can always be made into the transcendent. A transcendent society, however, has the negative effect of limiting human freedom. It relegates people to what Weber referred to as the iron cage of instrumental rationality and is thus something to be avoided. So the moderns were not incorrect in wanting an objective nature and a malleable and free society. "They were mistaken only in their certainty that double production required an absolute distinction between the two terms and continual repression of the work of mediation" (Latour 1993, 140). The second guarantee requires the introduction of a morality that demands that all practices that interfere with the "objectivization of nature" and the "subjectivization of society" be seen as impermissible and, again, immoral (Latour 1993, 140).

The third guarantee of the nonmodern constitution redefines freedom as the ability to sort associations without restraint (see Latour 1993, 141). Here, there are no modern temporal flows or cultural boundaries that restrict freedom of choice. The modern constitution restricted choices to the modern time frame and Western ideas. All objects, customs, and ideas that were of another time (or culture) were deemed useless and out of fashion. Because the moderns posited a great break between themselves and the past and others, they believed only what they had to say. All other ways of knowing, incorporating and mixing objects, and building associations were viewed as inferior and, as a consequence, something to be avoided or overcome. The nonmodern constitution attempts to overturn this tempocentrism and ethnocentrism. It allows for the mixing of things and people, past and present practices, and Western and non-Western modalities without undo legal regulation and restriction. Yet this guarantee is possible only if people are part of a malleable society. A transcendent society would make this freedom to mix objects and transform reality impossible.

The fourth guarantee of the nonmodern constitution argues that because hybrid production is now out in the open, knowledge making can be better controlled and regulated. Latour (1993, 141) contends that this allows for a more democratic approach to reality making. The modern constitution of truth granted to science a special origin myth and a secretive and extraordinary method. Science was said to be a vastly different way of knowing, void of the premodern or humanistic sins of speculation and belief. Science was also said to embody special techniques that only it could properly use and accurately report. Because science viewed itself as special and sacred, it essentially locked itself away from other realms, particularly anything of human origin. From this sacred position, society and unruly nature/society hybrids could only contaminate science's unbiased pursuit of reality and reporting of truth. The nonmodern constitution disallows these extraordinary aspects of science and calls for a rethinking of its need to be locked away. As a result, Latour thinks that it is now possible to have a discussion of the construction of reality without resorting to the defensive position of scientific privilege because that option is no longer of any use or value.

Latour's nonmodern constitution is an attempt to envision the interrelated aspects of all things that make up reality without resorting to ontological anchors. It calls for the recognition that it is actually quasi-objects—objects without any absolute essence—that are responsible for reality, not society or nature. In the actor-network approach "natures are present, but with their representatives, scientists who speak in their name. Societies are present, but with objects that have been serving as their ballast from time immemorial" (Latour 1993, 144). In this nonmodern constitution, science does not cease to exist. Rather, it is capable only of telling part of the story. Likewise, society does not cease to be of importance; rather, it too is only part of the story. Henceforth, nature will be the title given to those things that have been solidified and made transcendent and society will be the label given to things that are malleable and in the making.

Under the guarantees of Latour's nonmodern constitution, explaining knowledge moves from relegating objects to their proper society or nature poles—from providing and legitimating natural accounts of nonhumans and social accounts of humans—to tracing the practices of sorting and the networks that result from the mobilizations and stabilization of actants, both human and nonhuman (see Callon 1987, 99). Latour's approach, while no magic bullet for the contemporary crisis of epistemology (see Woolgar 1994), does attempt to restore the possibilities of the empirical while simultaneously freeing contemporary accounts of knowledge from an unending paradox.

To You Culture, To Us Nature: The Implications of the Modern Constitution of Truth and Its Demise

What bearing does the modern constitution of truth have on the contemporary debate among scientific realists, postmodern textualists, and social realists? In the broadest sense, the modern constitution of truth made possible the great battle cry of modern scientists: "to you culture, to us nature" (Latour 1990). It divided knowledge into distinct intellectual spheres and essentially guaranteed that "epistemology and political science will go their opposite ways" (Latour 1993, 29). The modern constitution of truth created three primary trajectories for twentieth-century knowledge. On one path, those on the natural side of the great divide set about the task of recording how all objects and events are products of nature. This first path finally culminated in a naturalization of all phenomena, as is found in such movements as "the theory of everything," sociobiology, or neuropsychology. On the second path, those on the social side of the divide set about the task of reporting how all objects and events are products of society or culture (see Akrich 1992). This second path finally culminated in the social-construction-of-everything approach found in social constructionism and the sociology of scientific knowledge. The third, more postmodern, path developed with the reflexive recognition that even knowledge about society, culture, or nature is itself textually constructed. This third path finally led to the textualization of everything as found in semiotics and deconstructionism.

On the first path, nature is made into a first principle from which all things ultimately derive. Truth is, therefore, merely the accurate reflection of that derivation. In the second case, as a "symmetrical counter claim," society and culture are made into first principles and the process is repeated on the other side of the divide (see Woolgar 1994, 9). On the social and cultural side of the great divide, however, truth has proven to be much more problematic. Scientists do not generally face spirally self-doubt about the potential truthfulness of their claims because their claims are seen as deriving from an ultimate and ontologically distinct "nature of things." They believe that this reality is beyond their ability to cognitively or linguistically change—nature exists before scientists encounter it. On the other side of the divide, however, social realists have faced reflexive doubts because they must at some point admit that they are enmeshed within the very object (i.e., society or language) from which their truth claims are derived. This reflexive turn creates the postmodern moment and its emphasis on textuality, rhetoric, and the indeterminacy of all truth claims. In the end, the modern constitution of truth leaves contemporary theories of knowledge with no other options than accurately reflected natural or social facts or socially and textually constructed cultural fictions.[3]

From this reading of the dilemmas posed by the modern way of knowing, it is possible to conclude that the debates now occurring in academia over the epistemological foundations of knowledge are the direct results of the divisions found in the modern constitution of truth. Scientific realists believe that it is their task to report the uncontaminated aspects of things in themselves. Those who disagree are either relativistic rhetoricians who are the sworn enemies of science and rationality or arrogant sociologists who, in their exuberance to reduce everything to social forces, display a lack of interest in or appreciation of the actual practices of those whom they study. In contrast, postmodernists believe that their assignment is to show how all reality is mediated by language and storytelling. Those who disagree are merely old-fashioned positivists whose scientism prohibits them from acknowledging the totalizing effects of rhetoric and language. Likewise, social realists believe that they must show how society is involved in every aspect of knowledge. Those who disagree are misguided naturalists who deny the all-consuming effects of the social or postmodernists who have gone completely relativistic and abandon truth. It can be argued that the modern constitution of truth has shaped the only current options available for those who want to understand and explain how knowledge works. It is, therefore, arguably responsible for the deadlock now so evident in the contemporary knowledge wars.

Notes

1. Latour's reading of the modern way of knowing owes much to the work of the philosopher Michel Serres. Serres's philosophy has a long history of denying the division between the modern and the nonmodern world. For example, Serres (1987) has equated the Challenger accident with the Carthaginians' incineration of people inside a brass representation of the god Baal. In his view, there is not a great break that separates the modern experience of the world from the premodern one: All events are temporally similar. All events involve people coordinating and managing their mythologies. Like Latour, Serres is also concerned with reconciling the divisions between science and society. He argues that science studies a world without people, while the social sciences study people without a world (see Serres 1995, 179).

2. The historical and ongoing "nature/nurture debate" can be seen as one of the key battles produced by the binary structure of the modern constitution of truth.

3. Latour points out, however, that the modern constitution is also flexible enough to allow for contradictory impulses even within the two poles of nature or society. For example, within the society pole, knowledge producers can be both relativistic and realist. Some can comment on the role of humans in crafting society (i.e., microinteractionists and praxis-oriented Marxists), while others explore the totalizing effects of social facts, norms, rules, and organization (i.e., structuralists and functionalists). Likewise, some within the natural pole can talk about capturing nature in the laboratory, while other can acknowledge the investigative craftwork needed in order to create and sustain objects in the laboratory.

Chapter 8

Conclusion: Scientific Realism, Social Realism, and Postmodernism Reconsidered

In the introduction I maintained that actor-network theory offers some new and innovative ways of reconfiguring truth. In this final chapter, I explore in greater detail that claim by examining some of the implications of actor-network theory for the positions of scientific realism, postmodern textualism and deconstructionism, and social realism. I also examine how actor-network theory figures into the current debate between modernists and postmodernists. Finally, I explore the ramifications of actor-network theory for the persistent and paradoxical issue of reflexivity found in both the sociology of knowledge and postmodern theories of knowledge.

In what follows, I first explore separately the implications of actor-network theory for scientific realism, postmodernism, and social realism before turning to the larger debate between modernists and postmodernists.

The Unrealities of Scientific Realism

There is no doubt that scientific realism holds a strong attraction. Scientific realism allows definitive truth claims to be made and evaluated. It settles the plaguing philosophical problems of epistemology and representation and the substantive predicaments of methodology in one grand sweep by grounding knowledge exclusively in the empirically observable and the logically derived. It is able to do this by setting up firm distinctions between the types of questions that can be answered empirically and experimentally and those that cannot and by neatly separating the knowing subject and knowable object. In addition, scientific realism promotes a social organization that is said to be both self-selecting and self-purifying—only honest and virtuous knowledge seekers need apply. In scientific realism one finds a strong conviction that humans can at long last comprehend and explain reality. From this view, science is the only method yet developed that properly overcomes the cultural and political intrusions that have always threaten to distort truth: Only science is able to sort myth from reality and defend reason from irrationality and superstition.

Scientific realism's disregard of the role of politics, culture, and history in shaping knowledge, however, leaves many dissatisfied and uneasy. For many critics of scientific realism, particularly feminist theorists, postmodernists, and sociologists of science, scientific realism's account of knowledge is hard to accept. When examining the history of science, it is relatively easy to find examples of alternative forms of knowledge that could have been but were not, or forms that still are capable of predicting and explaining but have now been relegated to the archives of bad ideas. When looking at politics in science, it is not difficult to find illustrations of partisan squabbles over intellectual property, such as access to funding sources and publication rights. When exploring cultural differences in knowledge, it is rather easy to find examples of varying forms of cultural representations that would signify phenomena equally well. These recognitions leave those concerned with assessing the validity of knowledge in a dilemma. On the one hand are the great successes of science that seem to confirm that it is indeed the most powerful strategy yet devised for speaking the language of reality. On the other hand are the conspicuous examples of science's embeddedness in culture, politics, and rhetoric that suggest that science is not that objective, context free, and representational after all.

One of the more obvious and plaguing problems of scientific realism to critics is that scientific knowledge seems to undergo constant change. "Science constantly changes its mind, not just with regard to incidentals but even on very fundamental issues" (Rescher 1987, 4). Scientific realists usually attribute this to the teleological nature of scientific progress. Science changes because it develops better, more truthful, and more representational accounts of reality. Contemporary science also claims, however, and must claim, that its current representations of reality are accurate. If science did not argue that its current representations were accurate, it would always be viewed as provisional and ineffective. It others words, if scientific views are forever tentative, we could never accept what science says about reality because that view might soon be wrong. Consequently, one of the key dilemmas for scientific realism is to maintain a belief in gradualism and progress while still defending the truthfulness of its current findings. It is this contradiction that, at least since the work of Kuhn (1962), makes scientific realism so untenable to critics.

So, what bearing does actor-network theory have on the doctrine of scientific realism and its reception and critique? Latour's work is often read by scientific realists as "an inadequate prop" for examining modern epistemology and as firmly antiscience (Gross and Levitt 1994, 62; see also Chalmers 1990, 80; Brown 1994). From the view of his critics, Latour's work is but another example of a faddish antirealism or an illustration of how inept sociologists really are in epistemological and scientific matters. Latour's approach does, however, attempt to offer a view of science that is void of both modern realist optimism and postmodern relativist pessimism. As such, it is neither a realist vindication of the progress of science or an antirealist denigration of all foundations. For example, Latour (1993, 142) writes:

> When we amend the Constitution, we continue to believe in the sciences, but instead of taking in their objectivity, their truth, their coldness, their extraterritoriality....we retain what has always been most interesting about them: their daring, their experimentation, their uncertainty, their warmth, their incongruous blend of hybrids, their crazy ability to reconstitute the social bond. We take away from them only the mystery of their birth and the danger their clandestiness posed to democracy.

In Latour's framework the plaguing question of whether science progresses or merely changes haphazardly becomes less central. The quest to explain the success of science need not be limited by an account of how scientific progress is possible. Such a quest is merely a result of the ontological limitations imposed on science and all of modern knowledge by the divisions and epistemic expectations of the modern constitution of truth. The demise of the modern constitution means that a new model of science can be developed that avoids the competing pulls of realism and relativism. In actor-network theory neither realism nor relativism can be used as explanations of scientific knowledge.

Latour seems to be offering a view of science that is both demythologizing and empowering. It is a view grounded in pragmatics rather than absolutist convictions. Science is to be recognized as an incredible achievement. It does produce wondrous findings and results. It still has foundations, although they are pragmatic and not epistemological ones. What actor-network theory drops is science's contention that it (and only it) captures and explains an independent reality. From such a reading, scientific realism turns out not to be as real after all, at least in the traditional sense advocated in scientific realism. That does not mean, however, that science is not to be admired. It only reconfigures the description of the process of how science is able to explain.

The Ironical Epistemological Vigilance of Postmodernism

From the previous chapters it is evident that postmodernism and actor-network theory do share some important similarities. Generally, they both have the desire to put an end to realist epistemology and accompanying correspondence theories of truth. Both postmodernism and actor-network theory reject the projects found in traditional epistemology, such as the quest to find a privileged vantage point from which to mirror reality and the argument that some distinct quality of the mind provides the cognitive foundation for the discovery of truth. Furthermore, both positions claim that the cherished modern epistemic distinctions between belief and knowledge, appearance and reality, and science and mythology no longer hold up under the weight of close deconstructive or empirical scrutiny.

As we have seen earlier, from the postmodern perspective, when we deconstruct modern epistemological distinctions we find a "violent hierarchy" rather than peacefully and naturally occurring demarcations (see Derrida 1981, 41). From the position of postmodernism, epistemic distinctions are linguistically biased and arbitrary and do not reflect real, ontologically derived opposition. From the perspective of actor-network theory, close observations of the daily workings of knowledge producers do not reveal adherence to the rigorous methodological rules of scientific procedures and truth finding, but instead finds mundane reasoners busy at work constructing and fortifying truth with various heterogenous materials. From this position, it is not the proper methodology or cognitive type that permits reality to be uncovered, but the successful construction of knowledge.

While these two antiessentialistic lines of thought obviously share a disbelief in all forms of foundational discourse, they are greatly divided as to what the much debated end of realist epistemology means for contemporary accounts of knowledge. If theory can no longer claim a superior position and its conceptual distinctions are arbitrary or culturally derived, what is the fate of scientific, social, or cultural theory? Often proponents of postmodernism argue that the demise of modern epistemology is the end of all forms of metanarratives or foundational discourses. This inevitably includes scientific metanarratives, one of the most important modern discursive formations. Since there is no correspondence between word and world, no theoretical position can possibly argue that it offers anything but narratives, texts, or chains of signs about the world. As such, the model of postmodern theory becomes a hermeneutically based literary criticism rather than scientific methodology. Within this perspective, theorists must become hermeneuticians who treat the world as a text to be continuously retranslated and reinterpreted. This unending hermeneutics does not seek to uncover or recover an ultimate meaning or reality, but rather merely to participate in the continuous cultural production of meaning.

It seems clear that both postmodernism and actor-network require theory to question the existence of philosophical dualisms. Postmodernism particularly questions the division between literature and science and signification and reality. Latour's approach asks that we question the long-held divisions between objectivism and relativism, nature and society, and humans and nonhumans, as well as the social and natural as ontologically distinct domains and foundations of explanations. This, however, is where the similarities between actor-network theory and postmodernism end. Latour is extremely hostile to postmodernism and its "hyper-incomprehensibility" (Latour 1993, 61). He (1990, 147) has described postmodernism as "the most sterile and boring intellectual movement to ever emerge" (Latour 1990, 147). This is so because from the position of actor-network theory, postmodernism contains a propensity to operate from within, rather than beyond, the dualisms of modern epistemology. In other words, "it is a symptom, not a fresh solution" (Latour 1993, 46). Latour (1990, 169) writes:

> The postmodern philosophers are not so happy, but they maintain the same structure (i.e., the modern structure). They are simply disappointed by the whole Critique enterprise and fail to believe anymore in the joint promises of rationalism and socialism. They have not moved an inch beyond. In spite of their presumptions this shows they are modern to the core.

It is in this respect that postmodernism can be said to display an ironic epistemological vigilance—realist epistemology may be dead, at least in different areas within the humanities and social sciences, but it continues to shape the options open for contemporary theorizing about knowledge. From the postmodern position, we have either objects captured by theory or fatal objects beyond representation, objectivism or relativism, reality or texts, rationality or irrationality, theory or poetics. It turns out that postmodernists, like their modernist counterparts, are also guilty of not allowing truth to vary—it is always either on or off. Only in the case of postmodernists, it is always off.

In making the point about the instability of representations, however, the postmodern position reveals an ultimately self-defeating paradox. Postmodernists reject traditional epistemology's search for foundations; however, they remain caught within epistemology's general philosophical vision and dualisms. In this sense, it can be argued that the conclusions of postmodern theory rest upon a false division contained within traditional epistemology and the modern constitution of truth. If we cannot have extrahuman knowledge, then *all* we can have are fictions, mythologies, texts, and a Baudrillardian "epistemology of predestination" (Game 1991, 167). Consequently, postmodernists "remain suspended between belief and doubt, waiting for the end of the millennium" (Latour 1993, 9). If word and world no longer correspond, then theory can become only poetically excessive or nihilistic (or perhaps silent). Theorists can only watch on the sidelines as the fatal destiny of the object unfolds.

In this textualizing move postmodernists do not acknowledge that they are making similar claims about truth—only this time in an inverted form. They are essentially playing by the rules and divisions put forth in the modern constitution of truth. Moving rhetoric and literature to the center does not essentially change the inherent logic of the modern order of things. It only elevates the once marginalized (e.g., literature, irrationality, rhetoric, and chance) to the position of the former (e.g., science, rationality, truth, and order). Like replacing patriarchy with matriarchy or the Occidental with the Oriental, it only switches the oppositions without reconfiguring or questioning the oppositions themselves. In retrospect, such a move proves to be quite contradictory since postmodernists have longed claimed that they are deconstructing the conceptual hierarchies of modernity (e.g., speech and writing, representation and reality, objectivism and relativism). Their deconstructionist textualism appears, however, not to have dismantled the important modern hierarchy between fact and fiction after all. In practice, postmodernists have only reversed it and transposed it into a new conceptual hierarchy.

From the position of actor-network theory, the existence of truth and reality are not, nor arguably have they ever been, dependent upon pure epistemically derived distinctions; nor do they just dissipate with their subsequent philosophical deconstruction and dissolution into text. Consequently, the loss of epistemic foundations is no need to mourn or to find a new label or alternative model for intellectual activity—although it does require a new model of knowledge. As Latour (1988b, 156) puts it, "denying rationality does not mean that the sky is going to fall on our heads, because the sky is supported by many other firmer pillars." These other pillars are associational in origin. Associational networks have a facticity beyond the reaches of idealist or textualist deconstruction alone. From this position, knowledge, truth, and reality do exist, but not as pure ideationally or philosophically distinguishable entities. Truth and reality are the cries of a strong coalition and a practically successful actor network. Truth and reality are (and have been) conditions that must be produced and continuously maintained through networks of actant activity. Literary critics who examine texts, nihilists who mourn the passing of foundations, or theoreticians who practice a hyper-guerrilla social criticism are merely parts of the other dimension of the modern constitution's objectivism/relativism dualism. As Latour (1988a, 169) has pointed out, "If all discourse appears to be equivalent, if there seems to be 'language games' and nothing more, then someone has been unconvincing." From this position, technoscientists or other knowledge producers do not sit in their laboratories and grieve over the loss of correspondence referentiality, begin the preparation to write geological poetry, or do subversive physics. They are actively engaged in a variety of practical actions and mobilization strategies that produce the real. The end of realist epistemology is not the end of knowledge generation, nor truth.

One of the most important implications of actor-network theory for postmodernists is its emphasis on the limits of textual deconstruction. From the position of actor-network theory, deconstructionists are only partly correct when they argue that science (or any form of knowledge) is built upon metanarratives, is symbolic, and uses rhetorical devices to constitute its knowledge and to sustain its division between the true and false. The deconstructionist approach to knowledge tells us little about how the constructions of science are achieved, maintained, and spread. Textual and symbolic structures are undoubtedly an important part of the overall construction of truth; however, they are not all or even the primary things that support scientific truth claims and the distinctions of science. Deconstructing the philosophical oppositions of "Big Science" (Price 1986) or displaying its rhetorical strategies does not in any way alter its ability to define the real, for science's power to nominate and speak the truth comes from an associationally constructed authority and not from the power of the words themselves.

Science is not the supreme modern way of knowing simply because of its philosophical, ideational, or rhetorical supremacy, but because it is a powerful associational network containing a strong and expansive web of heterogenous actants. It is capable of drawing on a vast array of allies, both human and nonhuman and textual and nontextual, in the construction and promotion of its truth claims (Latour 1987). In exposing the literary devices at play in science, deconstructionists leave intact the most important resource supporting science—its vast heterogenous network of human and nonhuman supporters. This is partly so because the ability to build large and encompassing networks of support is something usually not available to literary or nonscientific forms of knowing. As a result, literary or everyday statements are much more fragile than scientific ones. The distinctions between scientific and literary statements has less to do with the power of the words being spoken in the different knowledge-producing areas than the power of the networks doing the speaking.

It can be argued that in postmodern theory the Nietzschean equation of knowledge with power and rhetoric is worked out to its ultimate conclusion. In postmodernism, theorists must accept the fatal play and power of signification and practice poetry or other forms of nonrepresentational discourse. In Latour's actor-network approach we see the Durkheimian-initialed task of explaining knowledge in social terms taken in a new direction, allegedly void of the past transgressions of objectivism, relativism, reflexivism, and sociological reductionism. Here, theory must list and describe the circulation of actants as they go about the business of building networks of the truthful, rather than resorting to a foundational reality.

In a time and place when the old epistemological realism seems no longer tenable, thanks primarily to deconstructionism, neopragmatism, science and technology studies, and social constructionism, it becomes increasingly tempting to move to the other side—to relativize and textualize the world, including our own statements—to become Baudrillardian poets (see Gane 1991, 131). In the current postmodern climate, it is easy to see why everything becomes a semiotic object (Barthes 1976); why there are only representations– "a mobile army of metaphors, metonyms, and anthropomorphisms" (Nietzsche 1976, 46); why the attempt to reconcile social theory with social reality will always be "illusory and doomed to failure" (Baudrillard 1988, 99); why there is nothing outside of the text (Derrida 1976); why there appears to be only incommensurable language games rather than a central metanarrative (Lyotard 1984). Yet, as actor-network theory reminds us, these epistemic moves themselves are set up by an allegiance to the modern constitution of truth. All of these revelations are testimonies to the ironical epistemological vigilance of postmodernism. While it is true that postmodernists reject modern accounts of knowledge, they remain caught in modernism's basic dualisms between society and nature, human and nonhuman, and knowledge and ideology. In this sense, postmodernists do not solve the problems of relativism and reflexivity, they only intensify them and lead them to their ultimate nihilistic conclusions.

A Sociology without Social Realism

When taken together, Latour's attempt to dismantle the modern constitution of truth and his attempt to provide an amodern account of knowledge allow for a different look at the long-standing problems of representation and reflexivity in sociology and other disciplines. Sociology's primary task, since the incorporation of the modern constitution of truth into sociology by Durkheim and other nineteenth-century social theorists, has been primarily to provide social explanations of what has been purified to be social processes, except for sociological explanations that are somehow not as social. Sitting on the precipice between the natural sciences and humanities, social theorists have been forced to become Janus-faced moderns: realist one moment and relativist the next. As ironical but nevertheless diligent moderns, social theorists have been involved in the great purification process of modernity: "sorting out the kernels of science from the chaff of ideology" (Latour 1993, 35). They have been forced either to relegate other peoples' actions and beliefs to the realm of society and ideology by showing how everything is a social construction (while elevating their own to the transcendental truth); or, as with postmodernism and the reflexivity project, to treat all knowledge as symmetrical representations and face spiraling self-doubt about their own representations. In the first option, everyday truth claims are of the world, biased, mundane, mythical, untrustworthy, while sociologists are free floating, objective, sacred, scientific, and beyond refute. In this option, even when sociologists have gone ethnomethodological in order to examine first-order actor accounts, they have still retained a privileged position from which to interpret the constructed reality of those accounts. The ultimate price of this unrelenting and often times disguised realist sociology of knowledge, however, has been the arrival of relativism, irony, and reflexive contradiction. This move has undoubtedly opened the door to the second option now increasingly evident in sociology: the infinite deconstructionism, textualism, and reflexive constructionism of postmodernism.

The incongruity created by the poles of nature and society and the philosophical divisions between realism and relativism can be seen as being responsible for all the great overcoming strategies of twentieth-century sociology, from the Mannheimian sociology of knowledge to ethnomethodology to new literary forms. It can be argued, however, that these strategies have succeeded in intensifying the divisions between nature and society and knowledge and society rather than avoiding them. At a basic level, actor-network theory argues that "the constricting framework of sociological analysis" (Callon 1987, 100) has led social theorists to mistake "length or connection for differences in level" (Latour 1993, 120). The utilization of sociology's "pre-established social categories and rigid social/natural divide" (Callon 1987, 100) has forced sociologists to look at the order of things rather than at how things are ordered. These categories and divides have led sociologists to search for a foundational social reality rather than the network of the real. Sociologists have accepted the demarcation between humans and nonhumans, society and nature, and knowledge and ideology rather than looking at how networks do the sorting and purifying. They have looked at society as already always there rather than at the simultaneous construction of natures and societies in the process of the mobilization of actants.

Latour (1988b, 155) considers the problem of reflexivity that has historically haunted sociological accounts of knowledge to be a position from which one "cannot escape except by indefinite navel-gazing, dangerous solipsism, insanity and probably death." His approach seems to provide an avenue for coming to terms with the Durkheimian dilemmas described in chapter 4. In fact, these dilemmas can also be seen as results of the same modern constitution of truth. Within this legacy social theorists are unable to explain how their special knowledge is possible. They become caught in the position of either arguing for special status or trying to defend a radical relativism. It can be argued, however, that all of these moves take place under the direction and limitation of the modern constitution. Latour (1987, 256), however, rejects these responses:

> Our method would gain nothing in explaining 'natural' sciences by invoking 'social' sciences. There is not the slightest difference between the two, and they are both to be studied the same way. Neither of them should be believed more nor endowed with the mysterious power of jumping out of the networks it builds.

For actor network, using the knowledge of the social sciences to explain the knowledge of the natural sciences or any other knowledge form will always be self-defeating and counterproductive. "If nature and epistemology are not made of up of transhistoric entities, then neither are history and sociology" (Latour 1993, 27). This is so because society and knowledge are not separate entities but coproductions. Building knowledge is building society. "'Nature' and 'societies' are secreted as by-products of (the) circulation of quasi-objects" (Callon and Latour 1992, 349). Here, both society and knowledge (and nature) are the effects of the activity of knowledge producers and of all their human and nonhuman allies (Callon and Latour 1992, 350). Thus, sociology's object of study "is neither society itself nor so-called social relationships, but the very actor networks that simultaneously give rise to society and to technology" (Callon 1987, 99).

Based on the arguments of actor-network theory, we can conclude that reflexivity is a problem only within a regime that divided objects and events into the natural or the social and trapped and isolated sociological knowledge in the social domain. Once this regime and its distinctions are shown to be constructed rather than ontologically derived, it becomes possible to abandon the fascination with representation and, as a result, the problem of reflexivity. "Now if there is no immanence, if there are only networks, agents, actants, we cannot be disenchanted" (Latour 1993, 128). Within actor network's account of knowledge, sociologists, like all knowledge producers, are "fighting endlessly to create an asymmetry between claims, an asymmetry no one can reverse easily" (Latour 1987, 196). These sociological claims must proceed via networks, as do all other truth claims. They, too, must recruit allies, enroll others, choreograph objects, and resist deconstruction. In their radical reflexivity, sociologists should not abandon the hope of constructing truth, "because other storytellers seem to know and are constantly defining the actors that surround them" (Latour 1988a, 10). Latour (1987, 257) writes,

> The very definition of a society is the final outcome, in Sociology Departments, in Statistical Institutions, in journals, of other scientists busy at work gathering surveys, questionnaires, archives, records of all sorts, arguing together, publishing papers, organizing other meetings. The results of what society is made of do not spread more or faster than those of economics, topology or particle physics. These results too would die if they went outside the tiny networks so necessary for their survival.

Sociological statements, like all statements, can only ride on the tracks they build. The things called sociological knowledge or social theory are the cries of a strong coalition, not the accurate reflections of reality or the inaccurate representations of the real. This applies to Latour's own statements. In his words, "no explanation escapes the trials of strength that I have described—mine no more but also no less than another" (Latour 1988a, 148).

What does all this mean for the practice of sociology? In general, it requires that sociology abandon its rigorous adherence to the social side of the modern constitution of truth. It must consider nonhuman actants as cocausal with human actants and society and nature as coconstructions of actant mobilization. More specifically, actor network's approach to knowledge requires a transformation of sociology's mission from looking at how society causes knowledge or phenomena to looking at how both knowledge and society result from the mobilization and stabilization of actants. Within the modern constitution of truth, "Nature and Society allow explanation because they themselves do not have to be explained" (Latour 1993, 80). In actor-network theory, however, "the activity of nature/society making becomes the source from which societies and natures originate" (Latour 1992, 282). Society cannot produce nature nor can nature produce society, since both are crafted products of human and nonhuman actants (Callon 1987, Latour 1991). The thing we call nature and the thing we call society are the secreted by-products of the mobilization and circulation of actants (see Callon and Latour 1992; 349). The reason the exact sciences have eluded sociological analysis is "not because they are distant or separated from society, but because they revolutionize the very concept of society, of what it comprises" (Latour 1988a, 38). In this sense, nature work is always social work and vice versa; or, as Shapin (1988, 539) has put it, "solutions to the problem of knowledge are solutions to the problem of social order." Latour writes (1993, 6),

> Yes, the scientific facts are indeed constructed, but they cannot be reduced to the social dimension because this dimension is populated by objects mobilized to construct it. Yes, those objects are real but they look so much like social actors that they cannot be reduced to the reality "out there" invented by the philosophers of science.

In the end, knowledge and truth are merely the convenient labels we give to a coherent, stable, and encompassing network.

In sum, Latour wants a sociology that in some manner returns to its nineteenth-century definition—the study of associations. He wants to replace all distinctions, such as the sociological and philosophical ones between nature and culture, human and nonhuman, society and knowledge, and truth and falsehood with one: a division between weaker and stronger networks. In this new model, truth is allowed to vary. Here, truth is never static as it is in scientific realism, postmodernism, and social realism. It can always become stronger or weaker over time as actants congeal or disintegrate and as networks expand or contract. Such a reconceptualization of sociological truth is possible, however, only after the boundary distinctions and divides contained in the "modern constitution of truth" have been reworked.

Conclusion: A Final Comment on the Knowledge Wars

One of the potential strengths of actor-network theory's approach is that it adopts a view of truth, reality, and knowledge that is allegedly void of the objectivistic expectations of traditional realist epistemology, the relativistic conclusions of postmodern theory, and the reductionistic and reflexivity-ridden accounts found in social realism. As such, it holds the promise of keeping those concerned with understanding knowledge from becoming frustrated rationalists, either awaiting the coming or mourning the passing of correspondence referentiality, jubilant nihilists celebrating the absence of firm foundations, or maligned social realists trying desperately to defend their special sociological truth claims.

In arguing that knowledge is constructed via associational networks, Latour's empirical model of truth making tries to illustrate that we inhabit neither a place where word and world naturally correspond nor a banal place that is ad hoc, fatal, and ultimately meaningless. Rather, the world and its objects are to be reconceptualized as part of a complex, interrelated associational network. This network choreographs objects and events in such a way that the real is produced. The network enables actants to turn what were once disorganized opinions into ordered facts. Latour tries to convince us that building reality and truth, like building a freeway or a supercomputer, must be recognized as an intricately organized associational process. This truth and reality building is part of an ancient way of organizing objects and events that does not simply begin with an epistemic break between the modern and premodern or a radical Copernican paradigm shift.

Actor-network theory's amodern approach to knowledge points out the possibilities of maneuvering between the poles of objectivism and relativism, two of the results of the modern constitution of truth and its polarization and purification processes. In doing so, it seems to allow for a constructive view of knowledge that does not make the modern epistemological move of declaring that some truth claims are inherently better than others or the postmodern antiepistemological maneuver of declaring that all claims are inherently the same. As a result, it does not place contemporary theory in an either/or ontological dilemma as in the other approaches described in this book. At the epistemological level, Latour seeks to disassemble the modern order of things by dismantling the various epistemic and political distinctions of modernity, thereby providing the basis for an amodern way of knowing. Analytically and empirically, however, he is able to avoid the immediate reflexive question—of why we should accept his and actor network's version of knowledge formation as an accurate representation of how knowledge really works by arguing that all truth claims, including his own, require the mobilization of actants and the construction of networks. His views are not dependent on correspondence views of reality and, consequently, not plagued by the modern and postmodern spell of representation. Since all constructions must be carefully crafted in order to succeed, there is no need to seek special status for his claims, to employ special overcoming strategies, or to transform everything, including his own statements, into a semiotic text.

If one accepts actor network's account of knowledge, it is possible to conclude that the modern constitution of truth and the analytical and empirical models derived from it are largely responsible for the ontological, epistemological, and political positions found in the current knowledge wars in academia. In this case, one of the important contributions of actor-network theory is that it allows theories of knowledge to reconsider all of the positions and options contained within modernity, as well as side-step the modernism/postmodernism debate. In fact, the current knowledge wars between modernists and postmodernists can never be won. The modern constitution has a number of contradictory provisions that each side can take refuge in and utilize as ammunition for years to come. Neither side will ever be capable of disproving the other side's position because the modern constitution makes both positions equally tenable.

Despite all the various twentieth-century attempts in both social theory and philosophy to put an end to realist epistemology, its basic dualisms continue to occupy contemporary theories of knowledge. Today, however, both the modern and postmodern privileging of one side of these dualisms is becoming increasingly tiring and untenable. Many are convinced that regardless of hard we try we will never have representations of reality that transcend history or culture. We can never know reality-in-itself but can have only second-order texts. On the other side, many are convinced that postmodernism and its turn to rhetoric and unending interpretation is not the answer. These people are weary of reading and deconstructing their way to obliteration, of trying to ground truth in the infinitely expandable chorus of multiple voices, of metatheorizing their way to truth, of developing a stronger strong program, and of turning the world into a representation and then doubting their own representation. In this regard, Latour (1993, 115) writes:

> Haven't we shed enough tears over the disenchantment of the world? Haven't we frightened ourselves enough with the poor European who is thrust into a cold soulless cosmos, wandering on an inert planet in a world devoid of meaning? Haven't we shivered enough before the spectacle of the mechanized proletarian who is subject to the absolute domination of a mechanized capitalism and a Kafkaesque bureaucracy, abandoned smack in the middle of language games, lost in cement and formica?

In the interim, as the toll of the modern constitution of truth inflict itself on knowledge, the understanding of knowledge wallows in irony-inflicted self-doubt as the knowledge wars only intensify and the gulf between the two cultures widens.

Embedded with actor-network theory can be found a subtle call for a multidisciplinary approach to knowledge. The modern constitution of truth divided the ontological terrain into two zones and locked the pursuit of knowledge into discrete epistemological paths. As a result, it has spawned ferocious competition between practitioners within those zones and along these paths to show each other the true cause of phenomena. Because of this competition, those objects or people who strayed into the zone of another were summarily rounded up and returned to their proper nature or society domain. With the demise of the modern constitution and its ontological, epistemological, and political purification mechanisms, however, it is perhaps possible again to revive the call for a more encompassing and less restrictive view of knowledge. Certainly in academia today there are often calls for the development of more interdisciplinarity. These calls, however, usually lack conviction. The natural scientists, social scientists, and humanists, under the influence of the political divisions of the modern constitution, still conceive of each other as epistemic enemies. Like tribes who come into contact for the first time, those in the arts and humanities, social sciences, and natural sciences usually have difficulty communicating and vastly different expectations of knowledge. Once the epistemic divisions that promoted this division become no longer tenable, however, the ontological zones and political hierarchies can perhaps at last be reconfigured.

Finally, it is important to point out that while this book has been concerned with providing a constructive account of the contributions of actor-network theory and to situate it in the modernism and postmodernism debate and knowledge wars, this does mean that the position is beyond critique or analysis. There are undoubtedly many elements of actor-network's theory of knowledge that need alteration or fine tuning.[1] This may, however, be further evidence to illustrate that actor-network theory's truthfulness is in the making, not in its epistemic status. In this respect, it is perhaps most productive to see scientific realism, postmodernism, and social realism, as well as actor-network theory itself, as part of larger networks in the making. In the end, it is not necessarily a matter of which theoretical approach best describes how knowledge really occurs, the alternatives available for contemporary theorizing about knowledge, or who will ultimately win the knowledge wars. Such a stance would deny the postepistemological stance of actor-network theory. Whether the world will be seen as a definitive place where facts are found, a fatal realm where truth has withdrawn, a place where the organization of society is responsible for truth, or a constructive realm where truth is made will perhaps finally depend on the success of each of these positions to gain and maintain a network of truth.

Notes

1. There are a number of critiques that can be and have been made of actor-network theory. First, one could question the use of the metaphor of politics as a model for organizing reality. Is this politics of knowledge production an actual description of reality building or a metaphor? Second, it can be argued that Latour is, despite claims to the contrary, simply looking for a new starting point—in this case a "mega-foundationalism" (see Fuchs 1995). In this case, actor networks are merely another order of explanation that must, by necessity, maintain a form of realism. Third, as Collins and Yearley (1992) have argued, does actor-network theory provide only a way of reintroducing nature as the ultimate cause of phenomena? This time, however, it is done through the back door. In this case, actor-network theory is seen as being guilty of removing the social actor from the starting point of analyses of knowledge.

Bibliography of Works Cited

Agger, Ben. 1989. *Socio(onto)logy: A Disciplinary Reading*. Urbana, Ill.: University of Illinois Press.

—. 1993. *Gender, Culture and Power: Toward a Feminist Postmodern Critical Theory*. Evanston, Ill.: Northwestern University Press.

—. 1994. "Derrida for Sociology: A Comment on Fuchs and Ward." *American Sociological Review* 59: 501-05.

Agodi, Maria C. 1991. "Icaro Impigliato: L'ala Riflessiva della Sociologia Conoscenza Scientifica." *Sociologia-e-Ricerca-Sociale* 12: 34-62.

Aichele, George. 1994. "Literary Fantasy and Postmodern Theology." *Journal of the American Academy of Religion* 59: 323-37.

Akrich, M. 1992. "Beyond Social Construction of Technology: The Shaping of People and Things in the Innovation Process." In M. Dierkes and U. Hoffman, eds., *New Technology at the Outset: Social Forces in the Shaping of Technological Innovations, 173-190*. Boulder, Colo.: Westview Press.

Ashmore, Malcolm. 1989. *The Reflexive Thesis: Wrighting the Sociology of Scientific Knowledge*. Chicago: University of Chicago Press.

—. 1994. "Social Epistemology and Reflexivity: Two Versions of How to Be Really Useful." *Argumentation* 8: 157-61.

Atkins, Peter. 1995. "Science as Truth." *History of the Human Sciences* 8: 97-102.

Aune, James Arnt. 1990. "Rhetoric after Deconstruction." In R. Cherwitz, ed., *Rhetoric and Philosophy, 253-73* Hillsdale N.J.: Lawrence Erlbaum Associates.

Babich, Babette E. 1994. *Nietzsche's Philosophy of Science: Reflecting Science on the Ground of Art and Life*. Albany, N.Y.: SUNY Press.

Bacon, Francis. 1960. "Aphorisms on the Composition of the Primary History." In F.H. Anderson, ed., *The New Organon and Related Writings,* 273-84. Indianapolis: Bobbs-Merill Educational Publishing.

—. 1961a. *The Works of Francis Bacon, vol. 4,* ed. by J. Spedding, R.L. Ellis, and D.D. Heath. London: Longman.

—. 1961b. *The Works of Francis Bacon, vol. 6,* ed. by J. Spedding, R.L. Ellis, and D.D. Heath. London: Longman.

—. 1961c. *The Works of Francis Bacon, vol. 3,* ed. by J. Spedding, R.L. Ellis, and D.D. Heath. London: Longman.

Barnes, Barry. 1974. *Scientific Knowledge and Sociological Theory*. London: Routledge and Kegan Paul.

——. 1982. *T.S. Kuhn and Social Science*. London: Macmillan.
Barnes, Barry and David Edge, eds. 1982. *Science in Context*. Cambridge, Mass.: MIT Press.
Barnes, Barry, and David Bloor. 1982. "Relativism, Relationalism and the Sociology of Knowledge." In M. Hollis and S. Lukes, eds.; *Rationality and Relativism*, 21-47 Cambridge, Mass.: MIT Press.
Barthes, Roland. 1976. *Sade Fourier Loyola,* trans. by R. Miller. New York: Hill and Wang.
Baudrillard, Jean. 1983. *In the Shadow of the Silent Majorities, or the End of the Social.* New York: Semiotext(e).
——. 1988a. *Jean Baudrillard: Selected Writings*, ed. by M. Poster. Stanford, Calif.: Stanford University Press.
——. 1988b. *The Ecstasy of Communication.* New York: Semiotext(e).
——. 1991. "The Reality Gulf." *The Guardian* (January 11th).
Baynes, K., J. Bohman, and T. McCarthy, eds. 1987. *After Philosophy: End or Transformation.* Cambridge, Mass.: MIT Press.
Bazerman, Charles. 1988. *Shaping Written Knowledge: The Genre and Activity of the Experimental Article in Science.* Madison: University of Wisconsin Press.
Berger, Peter, and Thomas Luckmann. 1967. *The Social Construction of Reality*. New York: Anchor.
Best, Joel, ed. 1995. *Images of Issues: Typifying Contemporary Social Problems.* Hawthorne, N.Y.: Aldine de Gruyter.
Best, Steven. 1989. "The Commodification of Reality and the Reality of Commodification: Jean Baudrillard and Postmodernism." In J. Wilson, ed., *Current Perspectives in Social Theory*. Greenwich, Conn.: JAI Press.
Biagioli, Mario. 1993. *Galileo, Courtier: The Practice of Science in the Culture of Absolutism.* Chicago: University of Chicago Press.
Birke, L. 1986. *Women, Feminism and Biology: The Feminist Challenge*. New York: Methuen.
Blalock, Hubert. 1984. *Basic Dilemmas in the Social Sciences.* Beverly Hills: Sage.
Blanchot, M. 1986. "The Limits of Experience: Nihilism." In D. Allison, ed., *The New Nietzsche*. Cambridge, Mass.: MIT Press.
Bleier, R. 1984. *Science and Gender: A Critique of Biology and Its Theories on Women.* New York: Pergamon Press.
Bloor, David. 1976. *Knowledge and Social Imagery*. London: Routledge and Kegan Paul.
——. 1982. "Durkheim and Mauss Revisited: Classification and the Sociology of Knowledge." *Studies in History and Philosophy of Science* 13: 267-97.
——. 1983. *Wittgenstein: A Social Theory of Knowledge*. London: Macmillan.
Bourdieu, Pierre, Jean-Claude Chamboredon, and Jean-Claude Passeron. 1991. *The Craft of Sociology: Epistemological Preliminaries*, trans. by R. Nice. New York: Walther de Gruyter.
Boyd, Richard. 1984. "The Current Status of Scientific Realism." In J. Leplin, ed., *Scientific Realism*, 41-82. Berkeley: University of California Press.
Boyne, Roy, and Ali Rattansi, eds. 1990. *Postmodernism and Society*. New York: St. Martin's.
Briggs, John C. 1989. *Francis Bacon and the Rhetoric of Nature*. Cambridge, Mass.: Harvard University Press.

Brown, James R. 1994. *Smoke and Mirrors: How Science Reflects Reality*. London: Routledge.

Brown, Richard. 1987. *Society as Text: Essays on Rhetoric, Reason, and Reality.* Chicago: University of Chicago Press.

—. 1990. "Rhetoric, Textuality, and the Postmodern Turn." *Sociological Theory* 8: 188-98.

Callon, Michel. 1986. "Some Elements of a Sociology of Translation: Domestication of the Scallops and the Fishermen of St. Brieux Bay." In J. Law, ed., *Power, Action, and Belief: A New Sociology of Knowledge*, 196-229. Keele: Keele Sociological Review Monograph.

—. 1987. "Society in the Making: The Study of Technology as a Tool for Sociological Analysis." In W. Bijker, T. Hughes, and T. Pinch, eds., *The Social Construction of Technological Systems*, 83-103. Cambridge, Mass.: MIT Press.

Callon, Michel, and Bruno Latour. 1992. "Don't Throw the Baby Out with the Bath *School! A Reply to Collins and Yearley.* In A. Pickering, ed., *Science as Practice and Culture, 343-68*. Chicago: University of Chicago Press.

Carnap, Rudolf. 1967. *The Logical Structure of the World.* Berkeley, Calif.: University of California Press.

Cartwright, Nancy. 1983. *How the Laws of Physics Lie*. Oxford: Oxford University Press.

Chalmers, Alan. 1990. *Science and Its Fabrication*. Minneapolis: University of Minnesota Press.

Code, Lorraine. 1991. *What Can She Know? Feminist Theory and the Construction of Knowledge*. Ithaca, N.Y.: Cornell University Press.

Cole, Stephen. 1992. *Making Science: Between Nature and Society*. Cambridge, Mass.: Harvard University Press.

Collins, Harry, and Trevor Pinch. 1982. *Frames of Meaning: The Social Construction of Extraordinary Science.* London: Routledge.

Collins, Harry. 1981. "What is TRASP? The Radical Programme as a Methodological Imperative. *Philosophy of the Social Sciences* 11: 215-24.

—. 1983. "The Sociology of Scientific Knowledge: Studies of Contemporary Science." In *Annual Review of Sociology,* 256-85. Stanford: Stanford University Press.

—. 1985. *Changing Order: Replication and Induction in Scientific Practice*. London: Sage.

Collins, Harry, and Steve Yearley. 1992a. "Epistemological Chicken." In A. Pickering, ed., *Science as Practice and Culture*., 301-26. Chicago: University of Chicago Press.

—. 1992b. "Journey into Space." In A. Pickering, ed., *Science as Practice and Culture,* 369-89. Chicago: University of Chicago Press.

Condillac, Etienne de. 1980. *Logic.* New York: Abaris Books.

Condorcet, Marquis de. 1955. *Sketch for a Historical Picture of the Progress of the Human Mind* , trans. by J. Barraclough. London: Weidenfeld and Nicolson.

Culler, Jonathan. 1986. *Ferdinand de Saussure*. Ithaca, N.Y.: Cornell University Press.

Daly, Mary. 1978. *Gyn/Ecology: The Metaethics of Radical Feminism.* Boston: Beacon Press.

Davenport, Edward. 1987. "The New Politics of Knowledge: Rorty's Pragmatism and the Rhetoric of the Human Sciences." *Philosophy of the Social Sciences* 17: 377-94.

De Man, Paul. 1971. *Blindness and Insight: Studies in the Rhetoric of Contemporary Criticism*. New York: Oxford University Press.

Dentith, Simon. 1990. *A Rhetoric of the Real: Studies in Post-Enlightenment Writings from 1790 to the Present.* New York: St. Martin's Press.

Derrida, Jacques. 1976. *Of Gramatology,* trans. by G. Spivak. Baltimore: Johns Hopkins University Press.

—. 1981. *Positions* , trans. by A. Bass. Chicago: University of Chicago Press.

Descartes, Rene. 1952. R*ules for the Direction of the Mind.* Chicago: Encyclopedia Britannica.

—. 1988. *Descartes: Selected Philosophical Writings,* trans. by J. Cottingham, R. Stoothoff, and D. Murdoch. Cambridge: Cambridge University Press.

Douglas, Mary. 1975. *Implicit Meanings*. London: Routledge and Kegan Paul.

—. 1982. *Essays in the Sociology of Perception.* London: Routledge and Kegan Paul.

—. 1985. "Pascale's Great Wager." *L'Homme* 93: 1-30.

Dummett, Michael. 1978. *Truth and Other Enigmas*. Cambridge, Mass.: Harvard University Press.

Duran, Jane. 1991. *Toward a Feminist Epistemology*. Savage, Md.: Rowman and Littlefield.

Durkheim, Emile, and Marcel Mauss. 1963. *Primitive Classification,* trans. by R. Needham. Chicago: University of Chicago Press.

Durkheim, Emile. 1938. *The Rules of Sociological Method,* trans. by S. Solovay and J. Mueller. New York: The Free Press.

—. 1951. *Suicide*. New York: The Free Press.

—. 1953. *Sociology and Philosophy*. London: Cohen and West.

—. 1961. *Moral Education,* trans. by E.K. Wilson and H. Schnurer. Glencoe, Ill.: The Free Press.

—. 1965. *The Elementary Forms of the Religious Life,* trans. by J.W. Swain. London: Allen and Unwin.

—. 1973. "Sociology in France in the Nineteenth Century." In R. Bellah, ed., *Emile Durkheim on Morality and Society*, 3-24. Chicago: University of Chicago Press.

—. 1980. *Emile Durkheim: Contributions to l'Annee Sociologique*, ed. by Y. Nandan. New York: Free Press.

—. 1982a. *The Rules of Sociological Method,* trans. by W.D. Halls. New York: The Free Press.

—. 1982b [1909]. "The Contributions of Sociology to Psychology and Philosophy." In E. Durkheim's, *The Rules of Sociological Method*, 236-40. New York: The Free Press.

—. 1982c [1908]. "The Methods of Sociology." In E. Durkheim's, *The Rules of Sociological Method.* New York: The Free Press.

—. 1982d [1903]. "Sociology and the Social Sciences." In E. Durkheim's, *The Rules of Sociological Method, 175-210.* N.Y.: The Free Press.

—. 1983. *Pragmatism and Sociology*, trans. by J.C. Whitehouse. Cambridge: Cambridge University Press.

Featherstone, Mike. 1988. "In the Pursuit of Postmodern." *Theory, Culture & Society* 5: 195-215,

Fee, Elizabeth. 1983. "Women's Nature and Scientific Objectivity." In M. Lowe and R. Hubbard, eds., *Woman's Nature: Rationalizations of Inequality*, 9-28. Elmsford, N.Y.: Pergamon Press.

Findlay, Deborah. 1993. "The Good, the Normal and the Healthy: The Social Construction of Medical Knowledge about Women." *Canadian Journal of Sociology* 18: 115-35.

Fish, Stanley. 1989. *Doing What Comes Naturally*. Durham, N.C.: Duke University Press.

Fleck, Ludwik. 1979 [1935]. *Genesis and Development of a Scientific Fact,* trans. by F. Bradley and T.J. Trenn. Chicago: University of Chicago Press.

Foucault, Michel. 1970. *The Order of Things: An Archaeology of the Human Sciences*. London: Tavistock.

——. 1972. *The Archaeology of Knowledge,* trans. by A.M. Sheridan Smith. New York: Pantheon Books.

——. 1971. "Orders of Discourse." *Social Science Information* 10: 7-30.

——. 1979. "Power and Norm: Notes." In M. Morris and P. Patton, eds., *Michel Foucault: Power, Truth, Strategy*, 59-66. Sydney: Feral Publications.

Fraasen, C. van. 1980. *The Scientific Image*. Oxford: Clarendon Press.

Fraser, Nancy. 1993. "The French Derrideans: Politicizing Deconstruction or Deconstructing the Political?" In G. Madison, ed., *Working Through Derrida.*, 51-76. Evanston, Ill.: Northwestern University Press.

Fuchs, Stephan. 1992. *The Professional Quest for Truth: A Social Theory of Science and Knowledge*. Albany, N.Y.: SUNY Press.

——. 1995. "The New Wars of Truth: Conflicts Over Science Studies as Differential Modes of Observation." Unpublished paper: University of Virginia.

Fuchs, Stephan, and Steven Ward. 1994a. "What is Deconstructionism, and Where and When Does it Take Place? Making Facts in Science; Building Cases in Law." *American Sociological Review* 59: 481-500.

——. 1994b. "The Sociology and Paradoxes of Deconstructionism: A Reply to Agger." *American Sociological Review* 59: 506-10.

Fuller, Steve. 1993. *Philosophy, Rhetoric and the End of Knowledge*. Madison, Wis.: University of Wisconsin Press.

——. 1994. "The Reflexive Politics of Constructivism." *History of the Human Sciences* 7: 87-93.

Fyfe, G., and J. Law, eds. 1988. *Picturing Power: Visual Depiction and Social Relations*. London: Routledge.

Galison, Peter. 1987. *How Experiments End*. Chicago: University of Chicago Press.

Gane, Mike. 1988. *On Durkheim's Rules of Sociological Method*. London: Routledge.

——. 1991. *Baudrillard: Critical and Fatal Theory*. London: Routledge.

Garfinkel, Harold. 1967. *Studies in Ethnomethodology*. Englewood Cliffs, N.J.: Prentice Hall.

Geertz, Clifford. 1984. *Local Knowledge: Further Essays in Interpretive Anthropology*. New York: Basic Books.

Gergen, Kenneth. 1985. "The Social Constructionist Movement in Modern Psychology." *American Psychologist* 40: 266-75.

Giddens, Anthony. 1978. "Positivism and Its Critics." In T. Bottomore and R. Nisbet, eds., *A History of Sociological Analysis, 237-60.* New York: Basic Books.

Gieryn, Thomas. 1983. "Boundary-Work and the Demarcation of Science from Non-Science: Strains and Interests in Professional Ideologies of Scientists." *American Sociological Review* 48: 781-95.

——. 1982. "Durkheim's Sociology of Scientific Knowledge." *Journal of the History of the Behavioral Sciences* 18: 107-29.

Gilbert, Scott F. 1991. *Developmental Biology*, 3rd ed. Sunderland, Mass.: Sinauer Associates.

Gilbert, William. [1600] 1958. *De Magnete,* trans. by P. Mottelay. New York: Dover.

Glass, Bentley. 1960. "The Academic Scientist: 1940-60." *AAUP Bulletin* 46: 149-55.

Godlove, Terry F. 1986. "Epistemology in Durkheim's Elementary Forms of Religious Life." *Journal of the History of Philosophy* 24: 385-401.

Goffman, Erving. 1974. *Frame Analysis: An Essay on the Organization of Experience.* New York: Harper and Row.

Grafton, Anthony. 1991. *Defenders of the Text.* Cambridge, Mass.: Harvard University Press.

Gross, Alan. 1990. *The Rhetoric of Science.* Cambridge, Mass.: Harvard University Press.

——. 1991. "Rhetoric of Science without Constraints." *Rhetorica* 9: 283-99.

Gross, Paul, and Normal Levitt. 1994. *Higher Superstition: The Academic Left and Its Quarrels with Science.* Baltimore: Johns Hopkins University Press.

Gusfield, Joseph. 1976. "The Literary Rhetoric of Science." *American Sociological Review* 41: 16-34.

Habermas, Jurgen. 1979. *Communication and the Evolution of Society,* trans. by T. McCarthy. Boston: Beacon Press.

——. 1981. "Modernity versus Postmodernity." *New German Critique* 22: 3-14.

——. 1984. *The Theory of Communicative Action*, vol. 1, trans. by T. McCarthy. Boston: Beacon Press.

——. 1987. *The Philosophical Discourse of Modernity,* trans. by F. Lawrence. Cambridge, Mass.: MIT Press.

Hacking, Ian. 1983. *Representing and Intervening.* Cambridge: Cambridge University Press.

Haraway, Donna. 1989. *Primate Visions: Gender, Race and Nature in the World of Modern Science.* New York: Routledge.

——. 1991. *Simians, Cyborgs, and Women: The Reinvention of Nature.* New York: Routledge.

Harding, Sandra. 1986. *The Science Question in Feminism.* Ithaca, N.Y.: Cornell University Press.

——. 1989. "Feminist Justification Strategies." In A. Garry and M. Pearsall, eds., *Women, Knowledge,and Reality: Explorations in Feminist Philosophy*, 189-201. Boston: Unwin Hyman.

Hekman, Susan. 1990. *Gender and Knowledge: Elements of a Postmodern Feminism.* Cambridge: Polity Press.

Hesse, Mary, ed. 1976. *Revolutions and Reconstructions in Philosophy of Science.* Bloomington, Ind.: Indiana University Press.

Hull, David L. 1988. *Science as Process: An Evolutionary Account of the Social and Conceptual Development of Science.* Chicago: University of Chicago Press.

Hunter, Albert, ed. 1990. *The Rhetoric of Social Research.* New Brunswick, N.J.: Rutgers University Press.

Hutcheon, Linda. 1988. *A Poetics of Postmodernism: History, Theory, and Fiction.* New York: Routledge.

——. 1989. *A Politics of Postmodernism*. New York: Routledge.
Ijsseling, Samuel. 1976. *Rhetoric and Philosophy in Conflict*. The Hague: Martinus Nijhoff.
Jacobitti, Edmund. 1989. "Taking Humanism Seriously: Science and Rhetoric in the Postmodern World." *Differentia* 3: 91-118.
James, William [1907] 1991. *Pragmatism*. Buffalo, N.Y.: Prometheus.
Jameson, Fredric. 1972. *The Prison-House of Language: A Critical Account of Structuralism and Russian Formalism*. Princeton: Princeton University Press.
——. 1991. *Postmodernism, or, the Cultural Logic of Late Capitalism*. Durham, N.C.: Duke University Press.
Jardine, Lisa. 1974. *Francis Bacon: Discovery and the Art of Discourse*. Cambridge: Cambridge University Press.
Jones, Richard F. 1961. *Ancients and Moderns: A Study of the Rise of the Scientific Movement in Seventeenth-Century England.* St Louis, Mo.: Washington University Press.
——. 1971. "Science and English Prose Style, 1650-1675." In Stanley Fish, ed., *Seventeenth-Century Prose*, 441-501. New York: Oxford University Press.
Jones, W.T. 1952. *Kant and the Nineteenth Century.* New York: Norton.
Keat, Russell, and John Urry. 1982. *Social Theory as Science,* 2d ed. London: Routledge and Kegan Paul.
Keller, Evelyn Fox. 1982. "Feminism and Science." *Signs* 7: 589-602.
——. 1985. *Reflections on Gender and Science.* New Haven: Yale University Press.
Kellner, Douglas. 1988. "Postmodernism as Social Theory." *Theory, Culture and Society* 5: 239-69.
Kitcher, Philip. 1993. *The Advancement of Science: Science without Legend, Objectivity without Illusions.* New York: Oxford University Press.
Knorr-Cetina, Karin. 1981. *The Manufacture of Knowledge: An Essay on the Constructivist and Contextual Nature of Science.* Oxford: Pergamon.
——. 1994. "Primitive Classification and Postmodernity: Towards a Sociological Notion of Fiction." *Theory, Culture and Society* 11: 1-22.
Kolb, David. 1986. *The Critique of Pure Modernity: Hegel, Heidegger, and After.* Chicago: University of Chicago Press.
Kuhn, Thomas S. 1962. *The Structure of Scientific Revolutions.* Chicago: University of Chicago Press.
Lakatos, Imre, and Alan Musgrave, eds. 1970. *Criticism and the Growth of Knowledge.* Cambridge: Cambridge University Press.
Latour, Bruno, and Steve Woolgar. 1979. *Laboratory Life: The Social Construction of Scientific Facts*. Beverly Hills: Sage.
——. 1986. *Laboratory Life: The Construction of Scientific Facts,* 2d ed. Princeton: Princeton University Press.
Latour, Bruno. 1987. *Science in Action*. Cambridge, Mass.: Harvard University Press.
——. 1988a. *The Pasteurization of France*. Cambridge, Mass.: Harvard University Press.
——. 1988b. "The Politics of Explanation: An Alternative View. In S. Woolgar, ed., *Knowledge and Reflexivity: New Frontiers in the Sociology of Knowledge, 155-76.* Newbury Park, Calif.: Sage.
——. 1990. "Postmodern? No, Simply AModern! Steps Towards an Anthropology of Science." *Studies in the History of the Philosophy of Science* 21: 145-71.

——. 1991. "The Impact of Science Studies on Political Philosophy." *Science, Technology and Human Values* 16: 3-19.

——. 1992. "One More Turn after the Social Turn..." In E. McMullin, ed., *The Social Dimensions of Science*, 272-95. Notre Dame, Ind.: University of Notre Dame Press.

——. 1993. *We Have Never Been Modern*. Cambridge, Mass.: Harvard University Press.

Laudan, Larry. 1982. "A Note on Collins' Blend of Relativism and Empiricism." *Social Studies of Science* 12: 131-32.

Law, John. 1994. *Organizing Modernity*. Oxford: Blackwell.

Lawson, Hilary. 1985. *Reflexivity: The Post-Modern Predicament.* La Salle, Ill.: Open Court.

Layder, Derek. 1988. "Beyond Empiricism? The Promise of Realism." *Philosophy of Social Science* 15: 255-74.

Lehmann, Jennifer. 1993. *Deconstructing Durkheim: A Post-Post-Structuralist Critique.* London: Routledge.

Leplin, Jarrett, ed. 1984a. *Scientific Realism.* Berkeley, Calif.: University of California Press.

Leplin, Jarrett. 1984b. "Introduction." In his *Scientific Realism,* 1-7. Berkeley, Calif.: University of California Press.

Levi-Strauss, Claude. 1969. *The Raw and the Cooked.* New York: Harper and Row.

Lewin, Philip. 1994. "Categorization and the Narrative Structure of Science." *Philosophy and Rhetoric* 27: 35-62.

Locke, John. [1690] 1975. *An Essay Concerning Human Understanding,* ed. by P.H. Nidditch. Oxford: Clarendon Press.

Longino, Helen. 1990. *Science as Social Knowledge: Values and Objectivity in Scientific Inquiry*. Princeton: Princeton University Press.

Lorber, Judith. 1994. *Paradoxes of Gender.* New Haven: Yale University Press.

Lukes, Steven. 1972. *Emile Durkheim: His Life and Work.* New York: Harper and Row.

——. 1982. "Introduction." In E. Durkheim's, *The Rules of Sociological Method*, 1-27. New York: The Free Press.

Lynch, Michael, and Steve Woolgar, eds. 1990. *Representation in Scientific Practice.* Cambridge, Mass.: MIT Press.

Lynch, Michael. 1985. *Art and Artifact in Laboratory Science: A Study of Shop Work and Shop Talk in a Research Laboratory*. London: Routledge & Kegan Paul.

——. 1991. "Pictures of Nothing? Visual Construals in Social Theory." *Sociological Theory* 9: 1-21.

Lyotard, Jean-Francois. 1984. *The Postmodern Condition: A Report on Knowledge.* Minneapolis: University of Minnesota Press.

Madison, Gary, ed. 1993. *Working Through Derrida.* Evanston, Ill.: Northwestern University Press.

Mahsun, Carol Anne. 1987. *Pop Art and the Critics*. Ann Arbor, Mich.: UMI Research Press.

Mangena, Oshadi. 1994. "Against Fragmentation: The Need for Holism." In K. Lennon and M. Whitford, eds., *Knowing the Difference: Feminist Perspectives in Epistemology*, 275-82. London: Routledge.

Mannheim, Karl. 1936. *Ideology and Utopia,* trans. by L. Wirth and E. Shils. New York: Harcourt, Brace.

——. 1956. *Essays on the Sociology of Culture*. London: Routledge and Kegan Paul.

Manning, Philip. 1994. "Fuzzy Description: Discovery and Invention in Sociology." *History of the Human Sciences* 7: 117-23.

McCloskey, Donald. 1985. *The Rhetoric of Economics*. Madison: University of Wisconsin Press.

Merchant, Carolyn. 1980. *The Death of Nature: Women, Ecology and the Scientific Revolution*. New York: Harper & Row.

Merton, Robert K. 1938. "Science, Technology, and Society in Seventeenth Century England. *Osiris* 4: 360-632.

—. 1973. *The Sociology of Science: Theoretical and Empirical Investigations,* ed. by N. W. Storer. Chicago: University of Chicago Press.

Mestrovic, Stjepan. 1985. "Durkheim's Renovated Rationalism and the Idea that Collective Life is Only Made of Representations." *Current Perspectives in Social Theory* 6: 199-218.

—. 1988. *Emile Durkheim and the Reformation of Sociology.* Lanham, M.d.: Rowman and Littlefield.

—. 1991. *The Coming Fin De Siecle: An Application of Durkheim's Sociology to Modernity and Postmodernism*. London: Routledge.

Mulkay, Michael. 1979. *Science and the Sociology of Knowledge*. London: Allen & Unwin.

Murphy, Nancey. 1990. "Scientific Realism and Postmodern Philosophy." *British Journal for the Philosophy of Science* 41: 291-303.

Myers, Greg. 1990. *Writing Biology: Texts in the Social Construction of Scientific Knowledge*. Madison, Wis.: University of Wisconsin Press.

Nicholson, Linda, ed. 1990. *Feminism/Postmodernism*. New York: Routledge.

Nietzsche, Friedrich. 1955. *Beyond Good and Evil*. Chicago: Henry Regnery.

—. 1968. *The Will to Power,* trans. by W. Kaufmann and R.J. Hollingdale. New York: Vintage Books.

—. 1974. *The Gay Science,* trans. by W. Kaufmann. New York: Random House.

—. 1976. *The Portable Nietzsche,* ed. and trans. by W. Kaufmann. New York: Penguin.

—. 1989. *Friedrich Nietzsche on Rhetoric and Language,* trans. and ed. by S. Gilman, C. Blair, and D. Parent. Oxford: Oxford University Press.

Norris, Christopher. 1992. *Uncritical Theory: Postmodernism, Intellectuals and the Gulf War*. Amherst, Mass.: The University of Massachusetts Press.

Nye, D.A., and C.E. Ashworth. 1990. "Emile Durkheim: Was He a Nominalist or a Realist?" In P. Hamilton, ed., *Emile Durkheim: Critical Assessments*, 307-22. New York: Routledge.

O'Neill, John. 1995. *The Poverty of Postmodernism*. New York: Routledge.

Parker, Ian, and John Shotter, eds. 1990. *Deconstructing Social Psychology*. London: Routledge.

Peacocke, Arthur. 1984. *Intimations of Reality*. Notre Dame, Ind.: University of Notre Dame Press.

Pheby, Keith. 1988. *Interventions: Displacing the Metaphysical Subject.* Washington, D.C.: Maisonneuve Press.

Pickering, Andrew. 1984. *Constructing Quarks: A Sociological History of Particle Physics*. Chicago: University of Chicago Press.

Pickering, Andrew, ed. 1992. *Science as Practice and Culture.* Chicago: University of Chicago Press.

Pickering, Andrew. 1994. "The Mangle of Practice: Agency and Emergence in the Sociology of Science. *American Journal of Sociology* 99: 559-89.

Plato. 1966. *The Republic of Plato,* trans. by F. M. Cornford. New York: Oxford University Press.

Pollner, Melvin. 1991. "Left of Ethnomethodology: The Rise and Decline of Radical Reflexivity." *American Sociological Review* 56: 370-80.

Popper, Karl. 1959. *The Logic of Discovery.* London: Hutchinson and Co.

Porter, Theodore M. 1995. *Trust in Numbers: The Pursuit of Objectivity in Science and Public Life.* Princeton: Princeton University Press.

Portoghesi, Paolo. 1983. *Postmodern: The Architecture of Postindustrial Society.* New York: Rizzoli.

Prelli, Lawrence J. 1989. *A Rhetoric of Science: Inventing Scientific Discourse.* Columbia, S.C.: University of South Carolina Press.

Putnam, Hilary. 1975. *Philosophical Papers, vol. 1.* Cambridge: Cambridge University Press.

Quine, W.V.O. 1964. *Word and Object.* Cambridge, Mass.: MIT Press.

Reiss, Timothy. 1982. *The Discourse of Modernity.* Ithaca, N.Y.: Cornell University Press.

—. 1988. *The Uncertainty of Analysis: Problems of Truth, Meaning, and Culture.* Ithaca, N.Y.: Cornell University Press.

Rescher, Nicholas. 1987. *Scientific Realism: A Critical Reappraisal.* Boston: D. Reidel.

Restivo, Sal. 1990. "The Social Roots of Pure Mathematics." In S. Cozzens and T. Gieryn, eds., *Theories of Science in Society,* 120-43. Bloomington, Ind.: Indiana University Press.

Restivo, Sal, Jean Paul Van Bendegem, and Roland Fischer, eds. 1993. *Math Worlds.* Albany, N.Y.: SUNY Press.

Ritzer, George. 1991. *Metatheorizing in Sociology.* Lexington, Mass.: Lexington Press.

Rorty, Richard, ed. 1967. *The Linguistic Turn: Recent Essays in Philosophical Method.* Chicago: University of Chicago Press.

—. 1979. *Philosophy and the Mirror of Nature.* Princeton: Princeton University Press.

—. 1982. *Consequences of Pragmatism.* Minneapolis: University of Minnesota Press.

—. 1989. *Contingency, Irony, and Solidarity.* Cambridge: Cambridge University Press.

—. 1992. "Is Derrida a Transcendental Philosopher?" In D. Wood, ed., *Derrida: A Critical Reader,* 235-46. Oxford: Blackwell.

Rosser, Sue V. 1989. "Feminist Scholarship in the Sciences: Where are We Now and When Can We Expect a Theoretical Breakthrough?" In N. Tuana, ed., *Feminism and Science, 3-14.* Bloomington, Ind.: Indiana University Press.

Roth, Paul A. 1994. "What Does the Sociology of Scientific Knowledge Explain?: or, When Epistemological Chickens Come Home to Roost." *History of the Human Sciences* 7: 95-108.

Ruddick, Sara. 1989. *Maternal Thinking: Toward a Politics of Peace.* Boston: Beacon Press.

Rudwick, Martin. 1985. *The Great Devonian Controversy: The Shaping of Scientific Knowledge Among Gentlemanly Specialists.* Chicago: University of Chicago Press.

Sarup, Madan. 1989. *An Introductory Guide to Post-Structuralism and Postmodernism.* Athens, Ga.: The University of Georgia Press.

Schiebinger, Londa. 1993. *Nature's Body: Gender in the Making of Modern Science.* Boston: Beacon Press.

Schmaus, Warren. 1994. *Durkheim's Philosophy of Science and the Sociology of Knowledge*. Chicago: University of Chicago Press.

Schmid, Michael. 1982. "Habermas' Theory of Social Evolution." In J. Thompson and D. Held, eds., *Habermas: Critical Debates*, 162-80. Cambridge, Mass.: MIT Press.

Schutz, Alfred. 1967. *The Phenomenology of the Social World.* Evanston, Ill.: Northwestern University Press.

Sellars, Wilfrid. 1967. *Philosophical Perspectives.* Springfield, Ill.: C.C. Thomas.

Seidman, Steve. 1991. "The End of Sociological Theory: The Postmodern Hope." *Sociological Theory* 9: 131-46.

Serres, Michel. 1987. *Statues*. Paris: Bourin.

——. 1995. *Conversations on Science, Culture, and Time,* with B. Latour, trans. by R. Lapidus. Ann Arbor, Mich.: The University of Michigan Press.

Shapin, Steven, and Barry Barnes. 1977. "Science, Nature and Control: Interpreting Mechanics' Institutes." *Social Studies of Science* 7: 31-74.

Shapin, Steven, and Simon Schaffer. 1985. *Leviathan and the Air-Pump: Hobbes, Boyle and the Experimental Life*. Princeton: Princeton University Press.

Shapin, Steven. 1988. "Following Scientists Around." *Social Studies of Science* 18: 533-50.

——. 1994. *A Social History of Truth: Civility and Science in Seventeenth-Century England.* Chicago: University of Chicago Press.

Shapiro, Barbara. 1983. *Probability and Certainty in Seventeenth-Century England.* Princeton, N.J.: Princeton University Press.

Simons, Herbert, ed. 1990. *The Rhetorical Turn: Invention and Persuasion in the Conduct of Inquiry*. Chicago: University of Chicago Press.

Simons, Herbert. 1990. "Introduction: The Rhetoric of Inquiry as an Intellectual Movement." In H. Simons, ed., *The Rhetorical Turn: Invention and Persuasion in the Conduct of Inquiry*, 1-31. Chicago: University of Chicago Press.

Slaughter, M.M. 1982. *Universal Languages and Scientific Taxonomy in the Seventeenth Century*. Cambridge: Cambridge University Press.

Sloterdijk, Peter. 1987. *Critique of Cynical Reason,* trans. by M. Eldred. Minneapolis: University of Minnesota Press.

Snow, C.P. 1959. *The Two Cultures and the Scientific Revolution.* New York: Cambridge University Press.

Sprat, Thomas [1734] 1958. *History of the Royal Society,* ed. by J. Cope and H.W. Jones. St Louis: Washington University Press.

Tibbetts, Paul. 1986. "The Sociology of Scientific Knowledge: The Constructivist Thesis and Relativism." *Philosophy of the Social Sciences* 16: 39-57.

Traweek, Sharon. 1988. *Beamtimes and Lifetimes: The World of High Energy Physicists.* Cambridge, Mass.: Harvard University Press.

Turner, Bryan. 1994. *Orientalism, Postmodernism and Globalism.* London: Routledge.

Turner, Stephen, and Jonathan Turner. 1990. *The Impossible Science: An Institutional Analysis of American Sociology*. Newbury Park, Calif.: Sage.

Tyler, Stephen. 1987. *The Unspeakable: Discourse, Dialogue and Rhetoric in the Postmodern World.* Madison, Wis.: University of Wisconsin Press.

Tyndall, John. 1892. *New Fragments*. New York: D. Appleton.

——. 1905. *Fragments of Science, Part Two.* New York: P.F. Collier.

Vattimo, Gianni. 1988. *The End of Modernity: Nihilism and Hermeneutics in*

Postmodern Culture, trans. by J. Snyder. Baltimore: The Johns Hopkins University Press.

Vision, Gerald. 1988. *Modern Anti-Realism and Manufactured Truth.* London: Routledge.

Wallace, Walter L. 1995. "Why Sociology Doesn't Make Progress." *Sociological Forum* 10: 313-18.

Ward, Steven. 1993. "Constitution Intellectual Fields: Reductionism, Anti-Reductionism and the Social Organization of Economics and Sociology." *Michigan Sociological Review* 7: 60-79.

—. 1994. "In the Shadow of the Deconstructed Metanarratives: Baudrillard, Latour and the End of Realist Epistemology." *History of the Human Sciences* 7: 73-94.

—. 1995a. "The Making of Serious Speech: A Social Theory of Professional Discourse. *Current Perspectives in Social Theory* 15: 63-81.

—. 1995b. "The Revenge of the Humanities: Reality, Rhetoric, and the Politics of Postmodernism." *Sociological Perspectives* 38: 109-28.

—. 1996. "Postmodernism." *Contemporary Sociology* 25: 13-15.

—. Forthcoming. "Filling the World with Self-Esteem: A Social History of Truth-Making. *Canadian Journal of Sociology.*

Weber, Max. 1949. *The Methodology of the Social Sciences,* trans. by F. Kaufman. London: Oxford University Press.

—. 1958. *The Protestant Ethic and the Spirit of Capitalism,* trans. by T. Parsons. New York: Charles Scribner's Sons.

Westfall, Richard. 1958. *Science and Religion in Seventeenth-Century England.* New Haven: Yale University Press.

Whitney, Charles. 1986. *Francis Bacon and Modernity.* New Haven: Yale University Press.

Wilkins, John. 1675. *Of the Principles and Duties of Natural Religion.* London: A. Maxwell.

Woolgar, Steve, and Malcolm Ashmore. 1988. "The Next Step: An Introduction to the Reflexive Project." In S. Woolgar, ed., *Knowledge and Reflexivity: New Frontiers in the Sociology of Knowledge*, 1-11. London: Sage.

Woolgar, Steve. 1982. "Laboratory Studies: A Comment on the State of the Art." *Social Studies of Science* 12: 481-98.

—. 1988a. *Science: The Very Idea.* London: Tavistock.

—. 1988b. "Reflexivity is the Ethnographer of the Text." In S. Woolgar, ed., *Knowledge and Reflexivity: New Frontiers in the Sociology of Knowledge*, 14-34. London: Sage.

—. 1992. "Some Remarks about Positionism: A Reply to Collins and Yearley." In A. Pickering, ed. *Science as Practice and Culture*, 327-42. Chicago: University of Chicago Press.

—. 1994. "Science and Technology Studies and the Renewal of Social Theory." CRICT Discussion Paper #41, Brunel, the University of West London.

Woolgar, Steve, ed. 1988. *Knowledge and Reflexivity: New Frontiers in the Sociology of Knowledge.* London: Sage.

Yearley, Steve. 1988. "Settling Accounts: Accounts, Action and Sociological Explanation." *British Journal of Sociology* 39: 578-99.

Index

About the Author

Steven Ward is Assistant Professor of Sociology at Western Connecticut State University in Danbury, Connecticut. He is the author of several articles on postmodernism and the sociology of knowledge.